Municipal Solid Waste Management

Edited by Hosam El-Din Mostafa Saleh

Published in London, United Kingdom

IntechOpen

Supporting open minds since 2005

Municipal Solid Waste Management
http://dx.doi.org/10.5772/intechopen.79020
Edited by Hosam El-Din Mostafa Saleh

Contributors
Elsayed Elbeshbishy, Frances Okoye, Claudia Estela Saldaña Durán, Luciléia Granhen Tavares Colares, Aline Gomes de Mello de Oliveira, Gizene Luciana Pereira de Sales, Verônica Oliveira Figueiredo, Michael Addaney, Richard Kyere, Jonas Ayaribilla Akudugu, María Belén Almendro-Candel, Jose Navarro-Pedreño, Ignacio Gómez, Antonis Zorpas, Irene Voukkali, Pantelitsa Loizia, Muniyandi Balasubramanian, Iria Villar, Salustiano Mato, Akindayo Sowunmi, Boguslaw Bieda, Dariusz Sala, Aroloye Numbere, Hosam El-Din M. Saleh

First published in London, United Kingdom, 2019 by IntechOpen
IntechOpen is the global imprint of INTECHOPEN LIMITED, registered in England and Wales, registration number: 11086078, The Shard, 25th floor, 32 London Bridge Street
London, SE19SG – United Kingdom
Printed in Croatia

British Library Cataloguing-in-Publication Data
A catalogue record for this book is available from the British Library

Additional hard and PDF copies can be obtained from orders@intechopen.com

Municipal Solid Waste Management
Edited by Hosam El-Din Mostafa Saleh
p. cm.
Print ISBN 978-1-78923-831-0
Online ISBN 978-1-78923-832-7
eBook (PDF) ISBN 978-1-78985-632-3

For EU product safety concerns:
IN TECH d.o.o., Prolaz Marije Krucifikse Kozulić 3, 51000 Rijeka, Croatia, info@intechopen.com or visit our website at intechopen.com.

Meet the editor

Hosam Saleh is a professor of radioactive waste management in the Radioisotope Department, Nuclear Research Center, Atomic Energy Authority, Egypt. He has been awarded MSc and PhD degrees in Physical Chemistry from Cairo University. He is interested in studying innovative economic and environment-friendly techniques for the management of hazardous and radioactive wastes. Professor Saleh has authored many peer-reviewed scientific papers and chapters, and is the editor of different books from valuable international publishers. He serves as a reviewer and editor for several international journals. He was awarded the Scientific Encouragement Award from the Atomic Energy Authority (2013), the Encouragement Prize in Advanced Technical Sciences from the Academy of Scientific Research and Technology (2014), and was listed in *Marquis Who's Who in the World* for several editions.

Contents

Preface

This book provides guidance on the management, recycling, and disposal of municipal solid wastes. It refers to those wastes classified as aggregation of unwanted materials generated from a range of human-related activities denominated from domestic to production. These wastes are originated by human transit from migrant to settler modes of living, which impose the need to modify or change the character of raw or primary materials available to support or sustain new modes of living and originating human activity. The development and application of approaches and technologies that provide economic and safe management is an essential issue in the treatment and disposal of municipal solid wastes.

The authors collaborating in this project have summarized their experience and present advances in different fields related to assessing the management of these materials. The book contains 11 chapters, organized in four sections, that cover important research aspects in municipal solid waste management technologies. The first section consists of an introduction aimed at presenting a brief background to the generation, composting, types, and management of municipal solid waste.

The second section presents the management and recycling of municipal solid waste. It comprises four chapters that deal with the recycling of biowaste: the case of Pontevedra, prepared by Mato and Villar; using composted municipal solid waste under the concept of a circular economy and as a source of plant nutrients and pollutants, presented by Almendro-Candel et al.; management of organic solid waste in meal production, submitted by Colares et al.; and municipal solid waste management and inland water bodies, produced by Sowunmi.

Section 3 presents intelligent techniques for controlling municipal solid waste, where Boguslaw presents life cycle inventory modeling of municipal solid waste management systems in Kosodrza, a community of Ostrów; Saldaña Durán et al. prepared an urban management model: municipal solid waste for city sustainability; Kyere et al. submit decentralization and solid waste management in urbanizing Ghana; and Muniyandi presents household willingness to pay for improved solid waste management services.

The last section provides the topic of disposal of municipal solid waste with two chapters entitled: "Municipal solid waste disposal in mangrove forest: environmental implication and management strategies in the Niger Delta," submitted by Numbere, and "Improper disposal of household hazardous waste: landfill/municipal wastewater treatment plant," prepared by Elbeshbishy and Okoye.

The editor wishes to express his thanks to all participants in this book for their valuable contributions, and to Ms. Dajana Pemac for her assistance in finalizing the work. I also acknowledge all IntechOpen staff members responsible for the completion of this book and other publications for free visible knowledge.

Hosam El-Din Mostafa Saleh
Atomic Energy Authority of Egypt,
Cairo, Egypt

Section 1

Introduction

Chapter 1

Introductory Chapter: Municipal Solid Waste

Hosam M. Saleh and Martin Koller

1. Introduction

Rapid growth of the global population, permanently increasing life standards, and vast technological advancement are continually increasing the variety and amount of solid waste.

Generation of municipal solid waste, together with the high organic share present in solid waste and its often incorrect discarding, results in extensive ecological pollution, mainly based on the emission of gases that contribute to the greenhouse effect, such as methane (CH_4) and carbon dioxide (CO_2). Because of this environmental threat, municipal authorities are currently urged to implement techno-economic and political solutions of higher efficiency to manage the growing quantities of municipal solid waste [1].

The lion's share of municipal (mainly urban) solid waste consists of biodegradable matter, which plays a substantial role in greenhouse gas emissions in today's cities all around the globe. According to the present state of knowledge, integrated solid waste management is the strategy of choice to manage this issue; such strategies, however, require improvement in order to handle the growing organic fractions of municipal solid discards. If accomplished in a smart manner, this can on the one hand contribute to the aspired reduction of greenhouse gas emissions, and, on the other hand, even potentially generate economic benefits. Hence, systems for sustainable management of municipal solid waste are auspicious and attractive objects of study to assess current consumption behavior in different global regions and to protect the natural environment.

Generally, municipal solid waste gets disposed of in dumps and landfills as the most simple, convenient, inexpensive, and technologically less advanced method. Organic fractions as the major component of municipal solid waste undergo biodegradation under the anaerobic conditions prevailing in landfills, which consequently releases greenhouse gases as mentioned above [2].

Reduction or complete abolition of environmental contamination becomes increasingly important, which intensifies the global efforts dedicated to develop novel strategies for gradually reducing the quantities of the biodegradable municipal solid wastes in landfills. The process toward reduction of organic pollution involves (i) source separated collection of organic fraction of municipal solid waste, which undergo compost production, (ii) organic waste incineration for energy production, and (iii) mechanical/biological processing to get a compostable material [3].

This introduction chapter makes the reader familiar with the principles of municipal solid waste management, encompassing landfilling and recycling technologies; moreover, the composition of different types of municipal solid waste will be introduced. Based on this, the most feasible, promising, and realistic scenarios for municipal solid waste management are presented in order to provide a solid scientific background of these processes implemented or in development, and the factors needed to assess the sustainability of these processes in a critical and straightforward fashion by using innovative sustainable assessment tools [4].

2. Emergence and generation of municipal solid waste

"Municipal solid waste" is commonly understood as the waste accruing in a municipality. Most of this solid waste is generated without any segregation, and, therefore, it may be either harmful or harmless. In general, independent on the origin of municipal solid waste, its impact on the environment and different life forms affects pollution of air, water, and soil. Moreover, impact of municipal solid waste on land use, odors, and esthetic aspects has also accounted for holistic considerations of waste treatment systems.

In principle, the human species is on top of any environmental pollution and consequently constitutes the major factor endangering nature's biodiversity. Global population growth and increasing consumer demands, especially in strongly growing, emerging, and developing economies, have resulted in a large production increase worldwide. However, most industrial facilities have insufficient or completely lacking monitoring of their production processes in environmental terms, and often insufficient or inadequate facilities for management and treatment of waste. The global trend of rapid urban growth has further caused an increase of waste generation from private habitation sites and private and public service facilities; in addition, intensified construction and demolition activities are ongoing. As urban population density is generally very high all over the world, the daily consumption of goods and services is also high in urban areas. Additionally, the amounts of accruing municipal solid waste are also directly correlating with the economic status of the society in a given country [5].

Municipal solid waste generation *per capita* has increased in most of the countries globally; in many cases, this increase has been dramatic especially during the last years. Among all solid waste, plastics, paper, glass, and metals are the four categories of highest potential for recycling. The huge quantities of municipal solid waste are not only a severe ecological hazard but also cause major social concern. This makes it clear that appropriate municipal solid waste management is a current topic of utmost importance [6].

3. Composting of municipal solid waste

Because of diverse shortcomings such as the lack of waste segregation already at the origin, insufficient treatment, scarce reuse, lacking recycling systems, and often inappropriate disposal, solid waste management still has various gaps in the management chain which need to be filled. Treatment of the organic waste fraction for energy and resource recovery changes its physical and chemical characteristics. In this context, the most important processing techniques encompass composting (aerobic treatment) or bio-methanogenesis (anaerobic treatment in biogas reactors). Composting through aerobic processing produces compost as a stable product, which is broadly utilized as manure and as soil fertilizer and soil conditioner.

Due to various reasons, composting facilities are used to a lower extent in large metropolitan cities. Prevalence of unsegregated waste and production of low-quality compost resulting in low end user acceptance are the two most important reasons for this underutilization. Bio-methanogenesis via microbiological activity under anaerobic conditions generates biogas rich in methane as the value component. In general, composting becomes feasible when a given waste contains high moisture and high organic content. Uncontrolled and arbitrary disposal of mixed waste including organic fractions that cause environmental problems such as land pollution and pollution of soil and aquatic environments due to leaching of waste components [7].

An exemplary study assessing a new industrial process for mechanical-biological treatment of municipal solid waste reports that municipal solid waste received for treatment on the plant typically consists of, based on the dry mass, 9% of rejectable waste, 21% of fines (<20 mm) (mainly rejectables), 23% of paper and cardboard, and 15% of diverse plastic materials originating from petrochemistry. Such high content in plastics, paper, and cardboard is typical for the local situation (suburb of Mende, Lozère, France), where municipal solid waste is collected based only on a source separation of glass and complex residual waste, without separately collecting plastic, paper, and cardboards [8].

4. Types of municipal solid waste

A classification of solid waste sources can be accomplished based on the following assumptions:

i. All solid waste produced within a municipality's territory, independent on its physical and chemical nature and source of generation, is classified as "municipal solid waste" (**Figure 1**).

ii. All economic activities create a given solid waste pattern.

iii. Due to the fact that economic and consumers' activities cause generation of solid waste, all these activities are considered sources of solid waste [9].

Private households, hotels, offices, stores, educational, and other institutions are causes of municipal solid waste generation. The lion's share of solid waste encompasses organic (mainly food or horticulture) waste, cardboard, paper, plastics and other resins, textile rags, metal, and glass; in many cases, even demolition and construction debris is included in collected waste, in addition to certain quantities of precarious waste, such as batteries, electric light bulbs and fluorescent tubes, automotive parts, expired medicines and other pharmaceutical

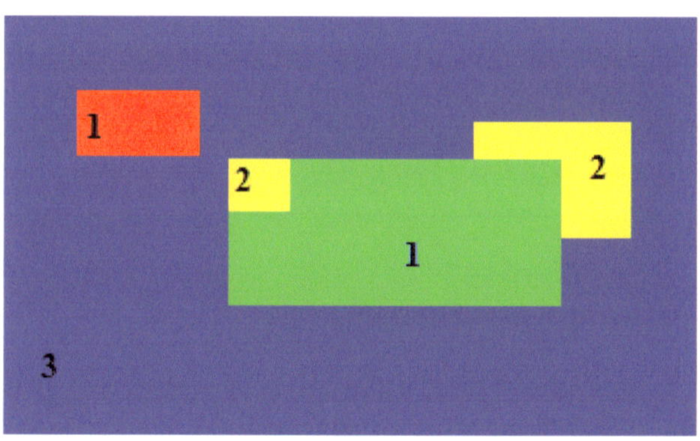

Figure 1.
A hypothetical urban municipality and the geographic areas (1. Urban, 2. Industrial, 3. Rural) where solid waste is generated [9].

Source	Typical waste generators	Types of solid wastes
Residential (private sector)	Single and multifamily habitations	Paper, cardboard, food wastes, plastics, textile rags, leather, yard waste, glass, lignocelluloses (wood, grass, and lopping), metals, ashes (heating and tobacco products), special wastes (e.g., bulky items, white goods, electronic parts, batteries, car tires, waste oils), and diverse types of precarious household waste
Industrial sector	Light and heavy manufacturing companies, fabrication, power and chemical plants, construction sites	Housekeeping waste, different packaging materials, food waste, construction and demolition materials, ashes, hazardous waste, and special waste
Commercial sector	Stores, markets, gastronomy, hotels, office buildings, etc.	Paper, cardboard, plastics, wood, food wastes, glass, metals, special wastes, and hazardous waste
Institutional sector	Schools, universities, kindergartens, hospitals and other health and medical institutions, penitentiaries, government centers	Same as for the commercial sector
Construction and demolition sector	New construction sites, renovation sites, road rehabilitation, demolition of buildings	Wood, steel, asphalt, cement, insulation materials, dirt, dust, etc.
Municipal services	Street cleaning, parks, landscaping, beaches, groves, playgrounds, sport facilities, other recreational areas, and wastewater treatment plants	Street sweepings, landscape, tree- and bush trimmings, different waste accruing in parks, beaches, riversides, and other recreational area, sludge after flooding events
Processing sector	Heavy and light manufacturing, chemical plants, (bio)refineries, power plants, mineral extraction and processing, joinery, and veneer works	Industrial process waste, saw dust, scrap materials, off specification products, slag, and tailings
All of the above should be included as "municipal solid waste"		
Agro-industrial sector	Farms, crops, orchards, vineyards, dairies, feedlots, distilleries, rendering and animal processing industry, biodiesel industry, and bioethanol production	Agricultural wastes, spoiled food wastes, animal residues (slaughterhouse waste), hazardous wastes (e.g., pesticides, antibiotic residues), and crude glycerol

Table 1.
Sources and types of solid wastes [10].

products, and diverse chemicals, e.g., cleaning and cosmetic products [10]. Hence, the main sources of solid waste are private households and the agricultural, industrial, construction, commercial, and institutional sectors. An assignment of different types of solid waste to their individual sources is shown in **Table 1**.

5. Municipal solid waste management

In parallel to the increase of population and economic activity, solid waste management is turning into a severe issue for almost all municipalities. Public health, odor disturbance, hazardous gas emissions, air pollution, or particulate matter formation are typical phenomena prevailing in urban regions. For smart management, municipal solid waste disposal requires proper environmental monitoring during

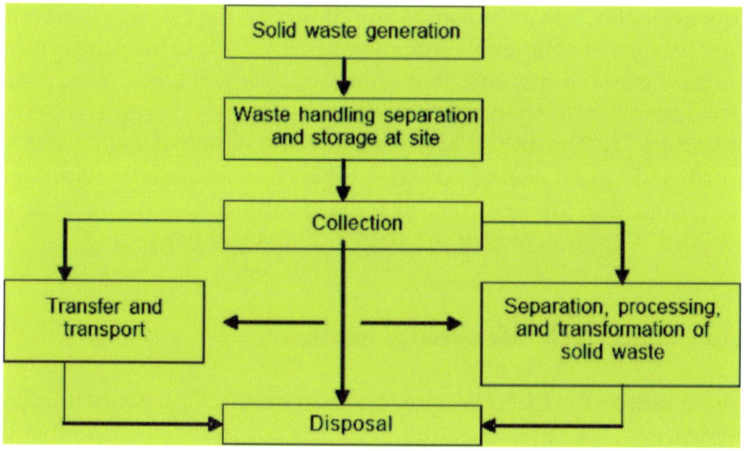

Figure 2.
Schematic of solid waste management system [12].

the entire waste treatment chain from waste collection to its ultimate disposal, and, finally, a regular control of disposal sites is needed [11].

To manage solid waste in an efficient fashion, the interrelationships of four functional elements have to be taken into account before a decision about an ultimate disposal strategy can be made. As reported by Shah [12], the first function element refers to the material generated at the source. Materials to which no more value is added are referred to and disposed as waste; quantity and nature of different types of waste are dependent on the waste source. The second function element encompassed the handling, separation, and storage at site of waste. In this context, waste has to be subjected toward separation before being placed into suitable storage containers. Paper, cardboard, packaging plastics, glass, ferrous metals, aluminum cans, and organic waste are those components, which typically are separated and stored individually. This step is crucial before moving to the next point. During the collection process, solid waste is picked up and placed into empty containers, which have separate compartments for recyclable materials [13]. Subsequently, the refuse collection staff collects the waste around the disposal centers manually before disposing it at the disposal sites. **Figure 2** illustrates the individual steps involved from waste material generation at its source until the final functional element for ultimate waste disposal.

6. Scenarios of municipal solid waste management

A policy for proper waste management needs to be grounded on the principles of sustainable development, which considers the society's refuse not only as rejects but also as a potential resource, which can undergo upgrading for potential value creation. In urban regions, appropriate solid waste management facilities are essential for, on the one hand, environmental management and protection and, on the other hand, for public health. Strategies and techniques for solving waste problems on a regional scale inevitably have a large number of possible solutions in order to be implemented in different areas, which are characterized by variable population densities, different life standard and life style, number of locations for waste management infrastructure, and number and types of protected landscape areas and other high value ecological sites. Environmentally benign waste management depends on various site-specific factors such as the composition of the waste, efficacy of waste collection at its source and of processing systems required to carry out different waste management techniques, feasibility of value-added material recovery from waste streams, emission standards to which waste management facilities are designed and operated, overall cost efficiency, and social performance of the community [7]. Due to this high complexity, municipal solid waste management has attracted a great deal of attention especially in countries with highly dynamic economic development such as India, a country that produces an estimated quantity of 50–600 million tons of municipal solid waste per year [7].

7. Municipal solid waste life cycle assessment

Life cycle assessment (LCA) is a process analytical tool recommended in many EU documents, e.g., the Directive 2008/98/EC on waste and certain other directives. LCA as a tool supports or enables the holistic consideration of the environmental impact of a new product or process already in its infancy, hence, during development [14]. As a quantitative measure, the Sustainable Process Index (SPI) allows to compare in a straightforward way the ecological footprint of products, processes, and systems based on the area required for completely embedding a process/system into the ecosphere [15]. Hence, LCA is a well-established tool, which nowadays is widely used to assess the environmental impact of product life cycles ("cradle-to-gate" or "cradle-to-grave"; the first refers only to production until the product leaving the factory's gate, while latter involves also the waste disposal after a product's life span), new technological processes, as well as waste management systems including waste treatment and processes for disposal, recycling, composting, or waste conversion for energy generation (biogas, thermal conversion in cogeneration plants). The evaluation of the existing situation of municipal solid waste management from an environmental, economic, and social perspective via a life cycle approach is an important first step prior to taking any decisions on the technologies to be selected, the policies to be developed, and the strategies to be followed for a nation [16].

The considerable number of reported LCA computer models dedicated to municipal solid waste management, often resorting to the SPI quantification tool, emphasizes the applicability of LCA in issues related to municipal solid waste management systems. Typically, these models have been developed independently from each other and are often based on features and assumptions that are highly specific to the period, economic framework, and geographical conditions in which they were developed. This clearly emphasizes that the assessment of feasibility of a given solid waste management systems needs to be in accordance to the individually prevailing conditions in a specific city or region.

Author details

Hosam M. Saleh[1]* and Martin Koller[2]

1 Radioisotope Department, Nuclear Research Center, Egyptian Atomic Energy Authority, Egypt

2 Institute of Chemistry, Office of Research Management and Service, University of Graz, NAWI Graz, Graz, Austria

*Address all correspondence to: hosamsaleh70@yahoo.com

References

[1] Abdel-Shafy HI, Mansour MSM. Solid waste issue: Sources, composition, disposal, recycling, and valorization. Egyptian Journal of Petroleum. December 2018;**27**(4):1275-1290

[2] Scaglia B, Salati S, Di A, Carrera A, Tambone F, Adani F. Short mechanical biological treatment of municipal solid waste allows landfill impact reduction saving waste energy content. Bioresource Technology. 2013;**143**:131-138. DOI: 10.1016/j.biortech.2013.05.051

[3] Scaglia B, Adani F. Sustainable landfill and biological stability of municipal solid waste. In: Lavelle JR, editor. Waste Management: Research, Technology and Developments. USA: Nova Science Publishers, Inc.; 2008. pp. 193-223

[4] Zhang DQ , Tan SK, Gersberg RM. Municipal solid waste management in China: Status, problems and challenges. Journal of Environmental Management. 2010;**91**:1623-1633

[5] Shekdar AV. Sustainable solid waste management: An integrated approach for Asian countries. Waste Management. 2009;**29**:1438-1448

[6] Karak T, Bhagat RM, Bhattacharyya P. Municipal solid waste generation, composition, and management: The world scenario. Critical Reviews in Environmental Science and Technology. 2012;**42**:1509-1630

[7] Ramachandra TV, Bharath HA, Kulkarni G, Han SS. Municipal solid waste: Generation, composition and GHG emissions in Bangalore, India. Renewable & Sustainable Energy Reviews. 2018;**82**:1122-1136

[8] Bayard R, de Araújo Morais J, Ducom G, Achour F, Rouez M, Gourdon R. Assessment of the effectiveness of an industrial unit of mechanical—Biological treatment of municipal solid waste. Journal of Hazardous Materials. 2010;**175**:23-32

[9] Buenrostro O, Bocco G, Cram S. Classification of sources of municipal solid wastes in developing countries. Resources, Conservation and Recycling. 2001;**32**:29-41. DOI: 10.1016/S0921-3449(00)00094-X

[10] Types of Wastes. World Bank; 1999. pp. 170-194. Available from: https://www.unescap.org/sites/default/files/CH08.PDF

[11] Chiemchaisri C, Juanga JP, Visvanathan C. Municipal solid waste management in Thailand and disposal emission inventory. Environmental Monitoring and Assessment. 2007;**135**:13-20

[12] Shah KL. Basics of Solid and Hazardous Waste Management Technology; Upper Saddle River, N.J.: Prentice Hall; 2000

[13] Tarmudi Z, Abdullah ML, Tap AOM. An overview of municipal solid wastes generation in Malaysia. Jurnal Teknologi. 2012;**51**:1-15

[14] Niederl-Schmidinger A, Narodoslawsky M. Life cycle assessment as an engineer's tool? Journal of Cleaner Production. 2008;**16**:245-252

[15] Narodoslawsky M, Krotscheck C. The sustainable process index (SPI): Evaluating processes according to environmental compatibility. Journal of Hazardous Materials. 1995;**41**:383-397

[16] Kulczycka J, Mineral T, Lelek L, Mineral T, Lewandowska A. Life cycle assessment of municipal solid waste management—Comparison of results using different LCA models. Polish Journal of Environmental Studies. 2016;**24**:125-140. DOI: 10.15244/pjoes/26960

Section 2

Management and Recycling of Municipal Solid Wastes

Chapter 2

Towards the Recycling of Bio-Waste: The Case of Pontevedra, Spain (REVITALIZA)

Salustiano Mato, Carlos Pérez-Losada, María Martínez-Abraldes and Iria Villar

Abstract

Waste management is one of the main environmental problems that municipalities have to address. The fulfilment of the recycling objectives imposed by the European Community requires the segregation and treatment of the municipal bio-waste. Pontevedra Provincial Council started in 2015 an innovative plan, called REVITALIZA, for the recycling of bio-waste through the promotion of composting in municipalities. REVITALIZA, which is developed in different phases, advocates the implementation of local composting (home and community composting) and small composting facilities, so that the generation of waste and the economic and environmental costs of its collection and transport are reduced. The plan is a pioneer in the training of technical personnel in the area of bio-waste management. Currently, 36 municipalities are participating in REVITALIZA in different phases of the plan, committed to locally managing bio-waste.

Keywords: composting, bio-waste, recycling, decentralized waste management, community composting centre, master composter

1. Introduction

Municipal waste generation in the European Union (EU) is estimated around 246,515 thousand tonnes in 2016, so the amount generated per person amounted to 483 kg [1]. Municipal waste represents only around 10% of total waste generated in the EU. However, its heterogeneous composition and universal distribution as well as the economic cost that the collection and treatment of this waste involve—especially for small and dispersed local authorities—cause a complex management and a high risk of environmental and socioeconomic impact in response to inadequate handling. Municipal waste management varies significantly across the EU member states. While Germany sent to landfill 2% and recycled and composted 66% of total municipal waste, countries such as Greece, Cyprus or Malta sent to landfill over 80% of municipal waste [1]. Municipal waste prevention and reuse, through responsible consumption, separation into the different elements found in waste streams and an appropriate management of these fractions, bring social and environmental benefits (priority hierarchy for solid waste management, [2]). The municipal waste management practices affect citizens: economic (waste collection fee) and environmental impacts (emissions and indirect system effects), but also more diffuse effects such as the physical connection with waste management through the design of the collection system and the psychological effect of the localization of waste management facilities [3]. The management of municipal waste must improve in order to move towards more sustainable systems, in accordance with the criteria of circular economy and with the involvement of citizens. European legislation and policy establish the necessary actions in order to ensure proper application of the waste hierarchy, turning waste into resources as a priority. The Waste Framework Directive sets a target of 50% of municipal waste to be prepared for reuse or recycled by 2020 in EU member states, progressively increasing this target up to 65% by weight by the year 2035 [2, 4]. Bio-waste, as part of municipal waste, is defined as biodegradable garden and park waste; food and kitchen waste from households, restaurants, caterers and retail premises and comparable waste from food processing plants. The content of bio-waste in municipal waste differs considerably between EU member states (20–80%); the implementation of separate collection and the treatment system of this biodegradable fraction also vary widely [5]. In general, by 31 December 2023, bio-waste must be either separated and recycled at source, or collected separately and not mixed with other types of waste [4]. Bio-waste is a valuable organic resource with a high potential for recycling and reuse, producing valuable products such as fertilizers or biogas. However, inefficient and neglected management can generate bio-waste breakdown and pollution, reducing the efficiency of subsequent treatment operations and generating human health and environmental impacts.

Around 20,585 thousand tonnes of municipal waste were generated in Spain in 2016, 57% sent to landfill and 30% recycled and composted, values far from the objectives established by the EU. Assuming that about 40% of municipal waste is bio-waste, this means that 8234 thousand tonnes of bio-waste were generated in Spain. In general, bio-waste is not source-separated and it is estimated that only 8% of this biodegradable waste is collected separately [6]. According to the 'Proximity Principle' of the EU, waste should be treated and disposed off close to where it was produced. Public administrations responsible for waste management should promote the local treatment of bio-waste. Food waste degrades quickly, generating leachates and odours, so reducing the waste treatment time prevents undesirable situations. Composting is an economically accessible and appropriate option, since it can be carried out at different scales with a simple and low-cost technology, which allows its location in places close to bio-waste producer. Composting is a controlled bio-oxidative process, which develops on heterogeneous organic substrates in solid state, due to the sequential activity of a great diversity of microorganisms. The process enables organic waste to be transformed into biologically stable materials called compost. The compost can be used as an amendment and/or soil fertilizer and as a substrate for plant growth, closing nutrient cycles. Due to the composition of municipal waste, source segregation of bio-waste at household level must ensure a material without other waste streams. Only the biodegradable fraction free from impurities can be used as input for composting to obtain a high-quality compost that is environmentally safe for use. Good waste segregation requires active participation of the citizens.

Actions that promote bio-waste segregation and the use of the most appropriate treatment options to obtain quality products, such as compost, must be prioritized in order to comply with current regulations and respect the environment. In this way, the implementation of new municipal models of organic waste management through composting is growing exponentially [7]. The decentralized management of biodegradable waste consists in on-site treatment (home composting, community composting and small composting facilities), while centralized management involves collection from the site of producer's deposit and transport to a central treatment facility.

In this sense, to change the centralized model of waste management in the province of Pontevedra, a composting plan is developed. This plan is based on the promotion and implementation of bio-waste composting with the criteria of population distribution and the prioritization of the principle of proximity in waste management. It includes an important effort in the awareness and training of citizens, as well as the training of professionals and experts in waste management in general, and the composting process in particular. The municipalities of the province of Pontevedra drive their efforts towards a decentralized model of bio-waste management.

2. The province of Pontevedra

Spain is made up of 50 provinces, one of which is Pontevedra. The province of Pontevedra lies in the northwest of the Iberian Peninsula. It is bordered to the south by Portugal and to the west by Atlantic Ocean (**Figure 1**).

The province has an area of 4495 km^2. It has a population of 942,731 inhabitants in 61 municipalities. With a population density of 209 inhabitants km^{-2}, Pontevedra is considered as an intermediate region according to the rurality indexes [8]. However, the population is more intensively concentrated in the metropolitan areas of the capital, Pontevedra, and the city of Vigo and along the coastal area. The eastern area of the province is a rural zone, with less densely populated municipalities, some of them with densities less than 50 inhabitants km^{-2}. **Table 1** shows the analysis of population dispersion data in the province. The population nuclei are distinguished according to the number of inhabitants in the settlement. A population nucleus, in its broadest sense, is considered to be a set of at least 10 buildings

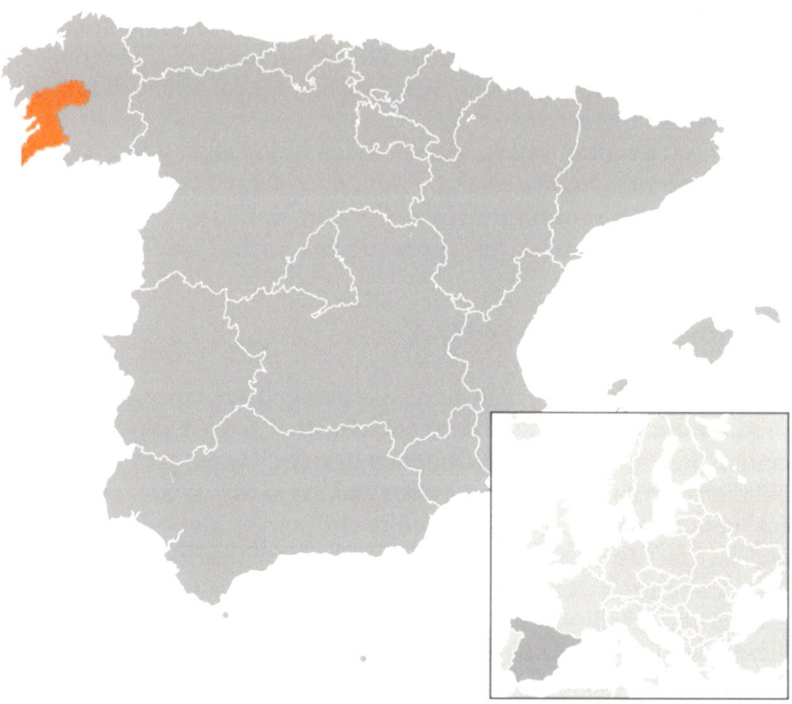

Figure 1.
Map of Spain showing its position in Europe; the study area of Pontevedra is marked with orange colour.

	Population nuclei				
	Scattered	<100	100–1000	>1000	Total
No. of inhabitants	209,918	69,175	197,732	484,179	961,004
No. of nuclei	17,493	1462	782	67	19,804

Table 1.
Dispersion data of the population of Pontevedra according to the size of the population nuclei. Source: prepared from the data of the Spanish Statistical Office [9].

that are made up of streets, squares and other urban roads; otherwise, the population is understood as disseminated. Around 30% of the population lives in small nuclei or scattered.

The dispersion of the population along with a complicated orography and a high rainfall involve a significant cost overrun in the public services, in general, and the municipal waste management in particular. In the waste management model of the province of Pontevedra, different waste streams are collected and managed separately: paper and cardboard, glass packaging and light packaging. Bio-waste is not source-segregated and it is collected in the mixed fraction, that is, all unsorted waste: bio-waste, sanitary textiles, ceramic waste, household cleaning waste, etc. The collection of these fractions takes place mainly in containers on public roads. Mixed fraction is managed in a centralized way in an incineration plant located at an average of 120 km from the municipalities of the province [7].

The services of collection, transport and treatment of the waste generated in the household, and similar sources such as commerce, offices and services, correspond to the municipalities. Each municipality decides how to provide and finance these services. These services suppose a high economic cost for the small and medium municipalities due to the difficulty in reaching a critical mass that optimizes the resources (containers, vehicles, staff, etc.) and the gap between the real cost of the services and the taxes applied to the citizens. Municipal waste generation in the province of Pontevedra accounted for 348,326 tonnes in 2017, but only 9.04% corresponded to separate waste collection, which led the municipalities far from the recycling objectives imposed by the EU.

The Provincial Council of Pontevedra is a supra-municipal authority, which provides direct services to citizens and technical, economic and technological support to the municipalities of the province of Pontevedra. The Provincial Council of Pontevedra has been promoting the composting plan called REVITALIZA since 2015.

3. Composting plan: 'REVITALIZA'

REVITALIZA establishes a new municipal waste management model focused on the segregation and treatment of the organic fraction as close as possible to its point of generation. It includes three fundamental bases or lines of action depending on the population distribution of the province and the particularities of each housing: individual or home composting, community composting and small composting facilities. The first two lines are considered as local composting, that is, composting near the area where the waste producers live. Small composting facilities should be located in the municipality or in a municipality near the places where the bio-waste is produced, and the waste will require collection and transport. **Table 2** presents a study of the theoretical requirements calculated for each one of the bio-waste treatment lines according to the population distribution of the province (**Table 1**). This study identifies what part of the bio-waste could be treated by local composting (home and community composting), while everything that could not be treated

	Population nuclei				Total	Population served (%)
	Scattered	<100	100–1000	>1000		
No. of home composters	51,980	16,794	11,796	6368	86,937	36.6
No. of CCCs	—	—	1230	1682	2912	36.7
No. of composting facilities	—	—	—	6	6	26.7

CCC: community composting centre.
Calculated following the assumptions: community composting centre of six units and small composting facilities of 3000 tonnes year^{-1} except one facility of 25,000 tonnes year^{-1} that would provide service to the city of Vigo.

Table 2.
Theoretical requirements of equipment and/or facilities for the implementation of the three lines of action (home composting, community composting and small composting facilities) of REVITALIZA based on population distribution data of the province of Pontevedra.

from a technical point of view through this priority path would be diverted to industrial composting at small and medium scale called small composting facilities.

As mentioned in the previous section, the population distribution of the province increases the cost of treatment of the waste, as there are small-sized nuclei and scattered population in dispersed areas. Based on the data in the table, most of the population of the province of Pontevedra can be served by local composting; the small facilities are restricted, especially for urban centres. Home composting is considered an interesting alternative to central composting, especially in areas with low population density [10]. According to REVITALIZA, the municipalities with scattered population would advance towards the sustainable management of the resources by means of local or in situ treatment of the bio-waste, so that municipalities would reduce the costs and environmental impacts of the collection and management of the mixed fraction of the municipal waste. REVITALIZA promotes the following actions:

- Establish a management model that allows to replicate and adapt it in the different municipalities of the province.
- Prioritize the treatment of the organic fraction near the point of generation and, therefore, reduce the collection and transport costs assumed by municipalities every day.
- Encourage the participation of citizens in waste management.
- Train personnel qualified in waste management, in general, and in the composting process, in particular.
- Obtain compost for use as a soil amendment and close the cycle of organic matter.

Initiatives to improve waste management services and the overall sustainability environmental policy chosen by local authorities require participation of all involved stakeholders (citizens, NGOs, state authorities, etc.). In order to be successful, all actions have to be credible, transparent, socially sustainable and, as far as possible, convenient and practical to participants [11]. Thus, consciousness-raising and training capacity for citizens are fundamental for the success of the composting plan. Therefore, experts on municipal waste management and composting process are required. The Provincial Council of Pontevedra has organized selection processes and specialized courses for the selection and training of staff called master composter. The courses had counted on the participation of expert teachers with recognized experience in the sector, both state and international. Master composters have as functions advising local governments on the composting plan and carrying out the actions, following the particularities of each municipality, for the implementation of REVITALIZA. In addition to the master composters, REVITALIZA has external collaboration from specialized associations: NGOs Amigos da Terra and ADEGA. These groups advise neighbours and control the operation of home composters. Likewise, personnel of the municipalities adhering to the plan will be trained so that they can take responsibility for the composting work in successive years. The Provincial Council of Pontevedra also carries out training actions addressed to the educational community, both teachers and students, through an agreement with the Center of University Extension and Environmental Outreach of Galicia, Spain (CEIDA).

3.1 Home composting

In accordance with the priority of minimizing the collection and transport of organic matter, the first level of REVITALIZA is local composting and, within it, individual or home composting.

The Provincial Council of Pontevedra transfers composters with capacity of 300 L to the houses with a plot of land (garden and orchard) (**Figure 2**). In this way, self-management of the bio-waste generated by the family nucleus can be carried out on site. In home composting, the participants segregate the bio-waste and deposit it in the composter, they are responsible for the composting process and they benefit from the obtained compost. Organic materials used for compost should include a mixture of food and kitchen waste and green organic material such as grass clippings, pruning remains, leaf litter, etc. Bulking agent is a carbon-based material such as chip or shredded pruning waste that creates necessary aeration structure for the composting process. The methodology used in home composters consists of alternating layers of food and kitchen waste with bulking agent that can be obtained in the garden of the participants themselves. REVITALIZA contemplates either the provision of bulking agent or the loan of crushers to process the garden waste from participants who require it.

The neighbours receive training and guidance from the associations that assist the Provincial Council. These associations carry out initial training and follow-up of the process through visits to each home composter. In the first year, at least three visits are made to check the development of the process: taking measurements of temperature, moisture control, filling level, incidents, etc. In addition, follow-up actions through telephone calls, emails, etc. are included.

3.2 Community composting

Community composting is a fundamental basis and strategic priority of REVITALIZA. Following the criterion of bio-waste management in areas close to the point of generation, composting at a community level consists in managing the bio-waste from local residents and/or activities within the same neighbourhood or community. To this end, community-composting centres (CCCs) are set up, either at neighbourhood communities or at small specific producers such as food stores, markets, bars, restaurants, hotels, etc. Community composting is considered an intermediate technique between home composting and composting in small-scale composting facility. In that sense, CCC that accepts more than 30 tonnes year^{-1} will be subject to specific legislation that includes, mainly, installations and environmental permit. A CCC is made up of modular units of 1 m^3 that serve around 20 inhabitants each. The minimum and maximum number of modular units per CCC

Figure 2.
(A) Home composter with an aerator on the lid, (B) material in composting process (C) bio-waste feeding to the composter and temperature measurement.

is 3 and 10 (**Figure 3**). The area of influence of a neighbourhood CCC is located at a maximum distance of 150 m from the homes it serves. In [12], it is observed that the larger is the distance of waste containers from the houses, the larger is the probability of waste dumping in other places. If the CCC is too far away, the probability that the neighbours deposit their waste in the container of the mixed fraction is greater. In the case of small producers' CCCs, they can be located in the producer's own facilities or in their proximity and must not exceed 30 tonnes year^{-1} of bio-waste.

Master composters evaluate the potential locations of the CCCs in the municipalities and the possible neighbourhood communities or small producers that would contribute bio-waste to the centres. The treatment capacity, the surface requirements and the material resources can be dimensioned according to the data collected by the master composters. An installation protocol has been developed for the placement of the CCC, in which the following criteria must be fulfilled:

- The land must be either municipal public property or expressly authorized by the owner.
- The land should preferably be natural and even with a maximum slope of 3%.
- The ground should be excavated about 20 cm deep for the installation of the base.
- The base consists of a lower layer of coarse gravel, a layer of fine gravel and the concrete pieces that make up the platform on which the modular units are seated and assembled.
- CCC must have a water feed for irrigation.

As an essential part of an appropriate composting process, food and kitchen waste must be mixed with bulking agent (crushed vegetable waste). This material is supplied by the municipality and comes from gardening activities, which involve pruning, cutting and removing vegetation of gardens, parks and other public spaces. The Provincial Council places at municipalities' disposal the crushing service, in case of lack of shredder equipment, so that they can prepare the remains of gardening to an optimum granulometric size for the community composting process [7]. Crates or bags with bulking agent are arranged in the CCC for use by the participants and master composters (**Figure 3**).

Participants of community composting receive initial training for the correct segregation and deposition of the bio-waste, as well as, information on the development of the composting process. The master composters continue their educational work in CCC on a day-to-day basis where they talk with the participants or interested parties and resolve their doubts and questions.

3.2.1 Composting process in CCC

The CCC working protocol is based on the complete development of the process in three modular units of composting: the first unit corresponds to the contribution or feeding module in which citizens deposit the bio-waste; while the second and third units are used to carry out the transfers (**Figure 4**). These transfers homogenize the material and, therefore, increase the efficiency of the process, which allows the first unit to be left empty for new contributions by the participants. Depending on the number of participants or the volume of bio-waste to be assumed, the number of modular units required in each CCC is set up. In this way, three stages are distinguished from the operational and process point of view in community composting.

3.2.1.1 Stage 1: bio-waste input

The neighbours deposit the bio-waste only in the modular units of feeding and immediately cover it with an equal volume of bulking agent. The master composters mix the materials so that the process begins. An intensive degradation phase takes place with a high oxygen demand, which is necessary for metabolic functions of the microorganisms. Large amounts of carbon dioxide and water vapour are released in this stage. The rise in temperature indicates that compost is developing properly.

3.2.1.2 Stage 2: homogenization

When the bio-waste input module is full, approximately in 4 weeks, the master composters move the material to the second module. The modular units are assembled together but have slide-out panels on all sides. This allows easy access on all sides and

Figure 3.
(A) Community composting centre with six modular units and bulking agent bags and (B) details of the modules during the composting process.

Figure 4.
(A) Bio-waste feeding in the first modular unit, (B) transfer of the material from the first module to the second to continue the composting process, (C) compost screening and (D) details of the physical aspect of the compost.

the movement of the material from one unit to another. Turning the material to the second module allows a more intense homogenization by mixing the most recent bio-waste inputs with degraded materials of the bottom. At this stage, the material might be too dry and the degradation process can stall; so, moisture control is important.

3.2.1.3 Stage 3: maturation

The material of the second unit is turned towards the third unit where the compost maturation takes place. The temperature drops progressively and more complex compounds are formed. The finished material has lost its original appearance. Compost is a soil-like material, dark with a pleasant earthy smell. Master composters sift the compost to facilitate its use as a fertilizer product or organic amendment. The compost can be distributed to citizens who have participated or can be employed by the municipal staff in the gardens and public areas.

As far as possible, the installation of urban or community gardens associated with CCC is promoted, so that the produced compost goes to the garden itself. The neighbours or small producers, instead of taking the compost produced for private use, would distribute the products of the garden. In [13], it has been proposed that the shift of municipal waste management systems from landfill disposal to resource recovery requires, among other aspects, sufficient urban gardens to divert the compost produced.

Throughout the process, master composters carry out the monitoring and control of composting and its key parameters (taking of temperature, filling level measurement, correction of incidents, etc.) and the necessary physical work required by the process (bulking agent addition, mixing, rewetting, turning, screening, etc.)

3.3 Small composting facility

The small composting facilities will manage bio-waste that cannot be treated through the other lines of action due to technical or operational causes. As has been described, local composting, both home and community composting, presents requirements for its implementation. In the case of high population densities distributed in buildings of various heights, local composting cannot assume all bio-waste generated. Therefore, it is necessary to implement a collection and transport service and bio-waste treatment in composting facilities. Following the principle of proximity, these facilities should be located close to the waste-production centres, so that the bio-waste transport is minimized and the treatment in areas near the point of generation is prioritized. These facilities must be small scale, handle between 1000 and 3000 tonnes year^{-1}. These will have limited mechanization given that the input waste cannot contain non-biodegradable materials or impurities (maximum allowed 10%). Medium-scale facilities could be established in the case of the two cities with the largest population of the province: Pontevedra and Vigo.

4. Development of REVITALIZA

4.1 Implementation phases

In order to ensure the success of REVITALIZA, its progressive implementation was considered necessary, so that the different actions will demonstrate the feasibility and effectiveness of the new model.

Authors present different possibilities:

- The first phases of the plan are assumed economically by the Provincial Council of Pontevedra.
- The Provincial Council of Pontevedra covers economically the implementation of the plan in the first phases.
- The Provincial Council of Pontevedra assumes the implementation of the plan in the first phases.

4.1.1 Phase I: demonstration stage

In order to demonstrate, both to citizens and public managers, the role of community composting in the province of Pontevedra, the Provincial Council put a selective process in motion at the end of 2015. This process was aimed at municipalities that were willing to implement the management of bio-waste through community composting. The municipalities interested in this new model were selected based on the following criteria:

- Submit agreements for the contribution of bio-waste free of non-biodegradable materials by neighbours and small producers to CCC.
- Keep a supply of bulking agent for mixing with the bio-waste.
- Pick up and use the compost by the neighbours or municipal services.
- Present adequate space available for the installation of the CCC.

From this announcement, 22 municipalities were selected and 221 modular units in 46 CCCs were installed in October 2016. In turn, master composters were selected and trained to give technical support, participate actively in the physical work of community composting and solve doubts and problems that may arise during the phases of the process.

DOI: http://dx.doi.org/10.5772/intechopen.83576

4.1.2 Phase II: adhesion and subsidies

Once the interest of the municipalities for community composting was demonstrated, the Provincial Council of Pontevedra started a new phase of REVITALIZA at the end of 2016. This phase consists in providing the necessary means for the implementation of the new management model as a global system for the treatment of bio-waste at the municipal level. With this objective, a second phase of REVITALIZA was established. Formally joining the plan was required for the municipalities to guarantee compliance with the legal obligations for the bio-waste treatment through composting. The formal adhesion of the municipalities allows them to benefit from three provincial collaboration lines: training of technical personnel, preparation of a municipal waste management plan and financial aid for composters and other resources supply.

This second phase has allowed the Provincial Council to begin the implementation of the new management model based on the local composting of bio-waste by home and community composting in five municipalities. These municipalities have decided to change the waste service betting on a decentralized model that will close the cycle of organic matter. These municipalities are Mondariz Balneario, Mondariz, As Neves, Vilaboa and O Grove. These municipal entities are small (between 1000 and 11,000 inhabitants) with a scattered population and few high-rise buildings. This new phase aims to manage 50% of the bio-waste produced in these municipalities through local composting in the next 2 years and reduce at least 25% of the organic fraction that is not reused (animal feed), donated (banks of food), composted or stabilized, within 4 years.

To give continuity to the plan and provide it with more personnel resources, different selective processes have been called and two training courses in composting have been carried out during 2017 and 2018.

4.2 Progress of REVITALIZA

Thirty-seven municipalities adhered to REVITALIZA, which represents 60.7% of the municipalities and 50.4% of the total population of the province. These municipalities are implementing the composting plan at different levels, either community composting or a municipal waste plan that includes home and community composting. The training and personnel selection activities have allowed 57 master composters who work at different levels and with different tasks and responsibilities. As part of the educational activities, 158 sessions were taught with 9448 participants, among students and teachers, in 51 educational centres.

The staff of the Provincial Council of Pontevedra actively participates in workshops, meetings, congresses, round tables, etc. that give visibility to REVITALIZA and allow to establish synergy with other institutions. REVITALIZA appears regularly in local and regional media reporting on the different events and activities that take place. These publications make it possible to give visibility to the plan not only at local and regional levels but also at national and international levels. Likewise, the neighbours and small producers who participate in composting serve as an example for the rest of the citizens, which allows to gradually involve more sectors of the municipality.

Next, the main results and advances of local composting are presented. Regarding the small composting facilities, the Provincial Council staff is making contacts with waste management companies with the aim of assuming the municipal bio-waste that cannot be managed by home and community composting.

4.2.1 Home composting

The first deliveries of individual composters started in the spring of 2018. As can be seen in **Table 3**, 37% of the composters expected delivery have been distributed in the five participating municipalities. The staff of the Provincial Council conducts door-to-door visits to collect data on the residents (address, number of family members, bio-waste management, etc.) in the areas of the municipalities that could manage the bio-waste by means of home composting. It should be pointed out that in more rural communities, traditional recovery of household waste at the household level, home composting and animal feed have diverted a part of bio-waste from municipal waste management system [14]. For this reason, a part of the rural population generates a small amount of bio-waste because of on-site reusing, to which one must add the second homes and the phenomenon of rural depopulation.

The master composters call the interested residents of the neighbourhood in which they are going to carry out the training and the delivery of composters. The training activities have been well received, with a percentage of attendance of 60% and an average of 33 composters delivered in 44 training activities.

During the follow-up visits to the home composters, the staff of the collaborating associations has solved doubts and established the necessary corrective measures related to the development of the composting process. The main incidents observed were the scarce quantity or lack of bulking agent and low moisture conditions of the composting material.

There is currently not enough data available to estimate the amount of bio-waste managed through this line of action. In [15], it has been estimated that in urban areas, where homeowners have access to garden space, home composting could potentially divert 20% of the biodegradable household waste stream from landfill disposal if approximately 20% of the community were actively engaged in home composting. The Provincial Council is studying the methodology to establish the amount of organic fraction treated in composters and, therefore, determine the contribution of home composting in bio-waste recycling.

4.2.2 Community composting

In November 2018, 76 CCCs formed by 535 modular units were in operation, spread over 28 municipalities in the province of Pontevedra. The master composters regularly visit CCCs, record the process parameters, such as temperature and changes in volume over time, and proceed to mix and turn the material between modular units, among other activities.

As previously mentioned, the work method allows the development of the composting process in three modular units. **Figure 5** shows the temperature profile of a

No. of home composters expected delivery	4186
No. of home composters delivered	1558
No. of training activities carried out	44
No. of home composters delivered/training activity	33
% of Participating neighbours/total neighbours called	60.0%
% of First visits/total home composters delivered	50.4%
% of Second visits/total home composters delivered	8.6%

Table 3.
Results of home composting in the five participating municipalities during 2018.

neighbourhood CCC and a small producer CCC, already settled in the population, during the monitoring of the process in the three units. The temperature profiles of both CCCs showed patterns typical of the composting process, in other words, temperatures increasing to thermophilic levels (>45°C) followed by maintaining said temperature and a subsequent decline in temperature until reaching mesophilic levels. Thermophilic temperatures were maintained for 65 days and 50 days in the neighbouhood CCC and small producer CCC, respectively. Despite these differences, compost hygienization was ensured by continuously maintaining temperatures above 55°C for more than 15 days [16]. In general, all CCCs reach high temperatures in the bio-waste input unit, although the development of the process will depend on numerous factors. Although, material in community composters are more isolated than the material present in home composters, the environmental changes can affect temperature development (periodic access for bio-waste input and for process control tasks). Another factor that affects the process is the type of bio-waste: uncooked and cooked waste. The biodegradation of recalcitrant compounds accelerates after the cooking process. On the other hand, [17] observed that when large amounts of waste were added at each feeding, compost temperature and maturity increased.

In the case of small producer CCC, it is observed that, after the turning of material from modular unit 1 (bio-waste input) to module 2 (homogenization), there was

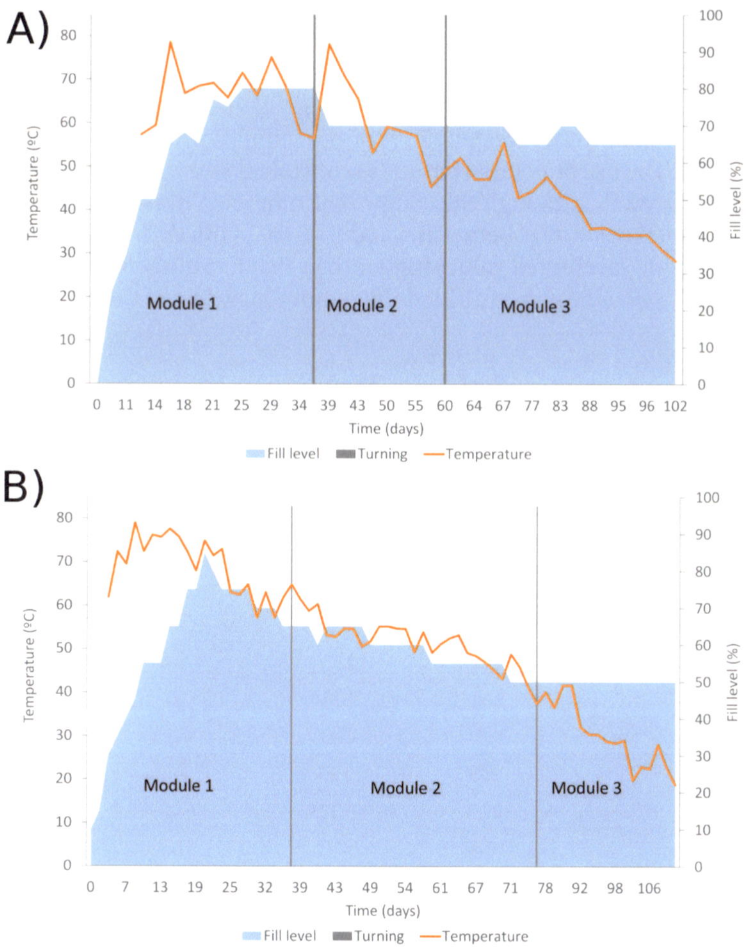

Figure 5.
Evolution of maximum temperature, fill level and turning during composting in the three modular units of (A) a neighbourhood CCC and (B) a small bio-waste producer CCC.

a rise in temperature. The master composters perform deep or superficial mixing of the composting material according to the conditions of the process. Although, the transfer of material from one module to another allows greater aeration and homogenization, which can facilitate an increase in temperature. In both temperature profiles, the phases or stages discussed above are distinguished: intensive degradation and temperature rise to thermophilic conditions (stage 1), less intense decomposition with maintenance and/or decrease in temperature (stage 2) and progressive decrease in temperature and maturation of the compost (stage 3). After the process, the compost presents homogeneous appearance (soil-like material), dark brown colour and a pleasant earthy smell (**Figure 4**). To facilitate the use of the product as fertilizer, potting soil or organic amendment, it is necessary to sift it.

Next, the analysis data of 76 composts sampled during the years 2017–2018 are presented (**Table 4**).

In general, composts showed high contents of organic matter, although the self-heating tests showed stability values indicative of mature compost. Important variabilities were observed among the compost for some parameters, such as electrical conductivity, ammonium and nutrients. The quality of municipal waste compost is dependent on many sources of variation including the composting facility design, feedstock source and proportions used, composting procedure, and length of maturation [18]. The different composition of the bio-waste affects the physicochemical characteristics of the compost. The high ammonium content could be a consequence of problems of degradation of the organic matter during the composting process due to a lack of moisture. However, only one sample had higher ammonium values than those considered suitable for compost 400 mg kg^{-1} [19]. Regarding electrical conductivity, high values were detected (78% of the samples with a conductivity higher than 2 dS m^{-1}). The use of compost must be controlled so as not to have negative effect on plant growth, although the compost of municipal waste usually presents electrical conductivity values between 4 and 8 dS m^{-1} [18]. As for pathogen content, 7.9% of the samples presented values higher than that established by the legislation for *E. coli* while *Salmonella* spp fulfilled the required level in all the samples.

	Mean	**Standard deviation**	**Legislation limit**	**Recommended values**
Moisture (%)	62.88	10.98	<40%	
Organic matter (%)	72.71	10.63	>35%	
pH	7.65	0.96	—	>7
Electrical conductivity (dS m^{-1})	3.39	2.32	—	<8
NH_4^+ (mg kg^{-1})	69.12	75.04	—	<400
CaO (%)	3.68	1.95	—	—
K_2O (%)	1.51	0.53	—	—
MgO (%)	0.43	0.49	—	—
P_2O_5 (%)	0.73	0.39	—	—
FeO (%)	0.38	0.28	—	—
Maturation degree	IV–V	—	—	IV–V
Salmonella spp (in 25 g)	Absence	—	Absence	—
Escherichia coli (CFU g^{-1})*	131.04	317.07	<1000 MPN	—

**N= 67 samples.*

Table 4.
Physicochemical parameters in compost from community composters (N = 76) during 2017 and 2018.

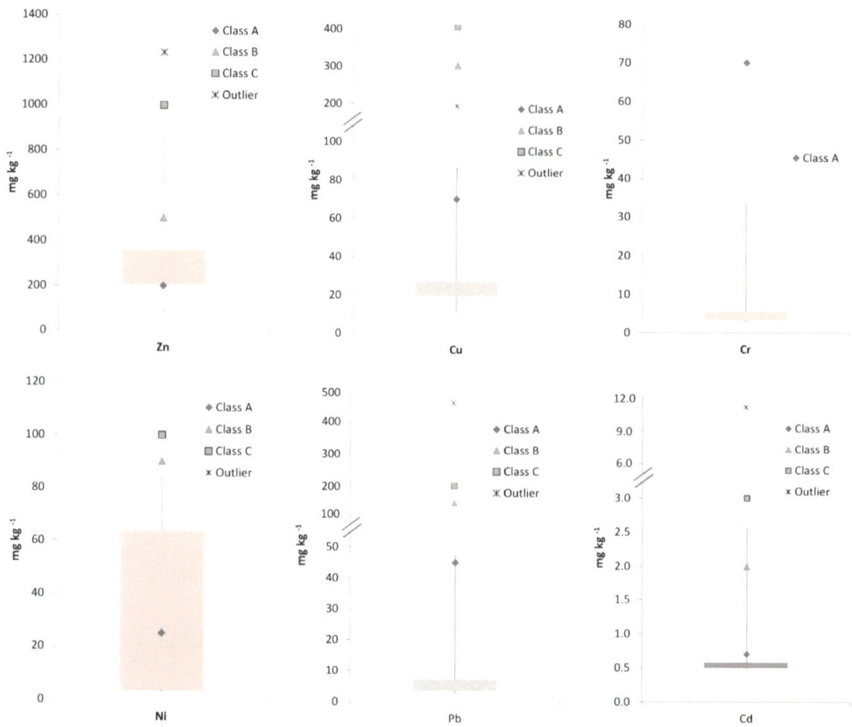

Figure 6.
Box plots of concentration data for six heavy metals in 76 samples of compost from community composting centres (CCC).

The Spanish legislation on compost [20] classifies compost into three categories according to the heavy metal content: Class A, B and C. **Figure 6** provides information on the variability in the heavy metals concentration indicating the respective classification categories. The atypical data observed for Zn, Pb and Cd correspond to different compost samples with a metal concentration 4 times (Zn), 7 times (Cu) and 52 times (Pb) above the mean values and, hence, they are considered outliers from analytical errors. Without taking into account samples with outliers, it is observed that 17.81% of compost belongs to Class A, 75.34% to Class B and 6.85% to Class C. For the last class, the metals Zn (4 samples) and Cd (1 sample) are those that exceed the thresholds of the regulations. There is a consensus in the scientific literature that aerobic composting processes increase the complexation of heavy metals in organic waste residuals and that metals are strongly bound to the compost matrix and organic matter, limiting their solubility and potential bioavailability in soil [21].

If we consider heavy metals separately, all samples belong to Class A for Hg (<0.4 mg kg^{-1} in all samples), Cr and Ni, while more than 96% of samples meet the levels for Class A in Cu and Pb concentrations. In 66.07% of Class B compost, Zn levels determine its classification. The presence of these heavy metals in the final compost may have different sources. In [22], it was concluded that the heavy metal content of the compost can be affected by the pollution of diverse exogenous sources and their origin can be found in the auxiliary materials used, the environment, the process or the storage method used. The possible sources of Zn are being evaluated to determine the necessary actions that reduce its content in the compost.

In **Figure 7**, the estimation of bio-waste treated in the CCCs of the province of Pontevedra is presented since the implantation of the first centres until the first semester of the year 2018. The quantities of treated bio-waste were calculated from the data of filling level of the CCCs, percentage of volume reduction over time

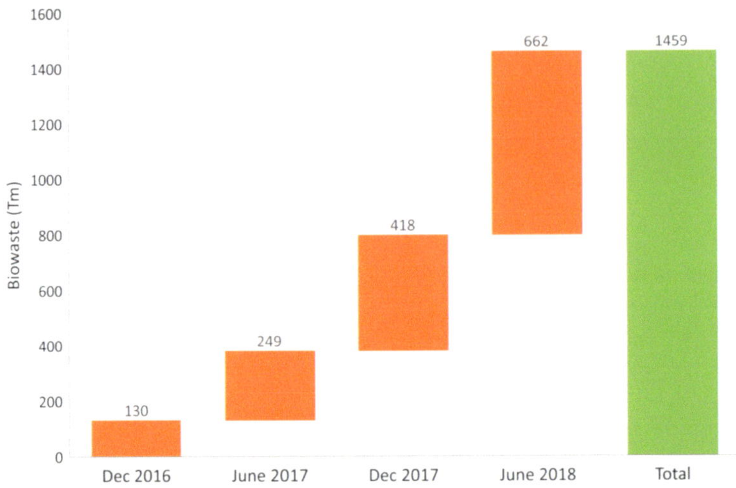

Figure 7.
Estimation by semester of the amount of bio-waste (including the vegetal fraction used as bulking agent) and the total amount accumulated in the CCC implemented in the province of Pontevedra.

and densities of the different materials. The bulking agent: food waste ratio 1:1 in volume was considered.

Finally, it should be noted that community composting have transformed, through a biological and aerobic process, about 1459 tonnes of organic waste and vegetable remains, into a biologically stable material that can be used as a soil amendment. This reduces the impact of bio-waste on the environment and makes possible the use of the resources that it contains.

5. Conclusions

The Provincial Council of Pontevedra promotes a change of model of waste management through the implementation of composting as treatment of the organic fraction generated in the municipalities, reducing the collection and transport services and the environmental and economic problems associated with them.

The new model has been designed to respond to the particularities of the province and the municipalities that compose it, so that it adapts to the population distribution characterized by dispersion in rural areas. This fact, together with the priority of compliance with the principle of proximity in the waste management, has made it possible to move towards a decentralized model based on the local composting of bio-waste at the municipal level. The provision of personal resources, and not only material resources, presented by REVITALIZA is a fundamental and necessary axis that demonstrates that the waste management projects developed by the administrative entities must be accompanied by training and raising of awareness to be accepted by the citizens.

Local composting allows the treatment of the bio-waste of the household and small producers on site. Bio-waste ceases to be part of the collection, transport and treatment line of the mixed fraction, thus reducing the environmental implications caused by its centralized management.

Acknowledgements

The authors thank the staff of the Provincial Council of Pontevedra for the technical support and the data provided.

DOI: http://dx.doi.org/10.5772/intechopen.83576

Author details

Salustiano Mato[1], Carlos Pérez-Losada[2], María Martínez-Abraldes[2] and Iria Villar[1]*

1 Department of Ecology and Animal Biology, University of Vigo, Vigo, Spain

2 Pontevedra Provincial Council, Pontevedra, Spain

*Address all correspondence to: iriavillar@uvigo.es

References

[1] Eurostats. Municipal Waste Statistics [Internet]. 2016. Available from: https://ec.europa.eu/eurostat/statistics-explained/index.php/Municipal_waste_statistics#Municipal_waste_generation

[2] Directive 2008/98/EC of the European Parliament and of the Council of 19 November 2008 on Waste and Repealing Certain Directives [Internet]. No. 312, 32008L0098. 2008. Available from: http://data.europa.eu/eli/dir/2008/98/oj/eng

[3] Reich MC. Economic assessment of municipal waste management systems—case studies using a combination of life cycle assessment (LCA) and life cycle costing (LCC). Journal of Cleaner Production. 2005;**13**(3):253-263. DOI: 10.1016/j.jclepro.2004.02.015

[4] Directive (EU) 2018/851 of the European Parliament and of the Council of 30 May 2018 Amending Directive 2008/98/EC on Waste [Internet]. No. 150, 32018L0851. 2018. Available from: http://data.europa.eu/eli/dir/2018/851/oj/eng

[5] European Environment Agency. Managing Municipal Solid Waste—A Review of Achievements in 32 European countries. EAA Report No. 2/2013. Copenhagen; 2013. DOI: 10.2800/71424. ISSN: 1725-9177

[6] Fundación Conama. Documento del Grupo de Trabajo de Conama 2016: GT15. El desafío de la gestión de la materia orgánica [Internet]. 2016. Available from: http://www.conama.org/conama/download/files/conama2016//GTs%202016/15_final.pdf

[7] Villar I, Alves D, Mato S, Romero XM, Varela B. Decentralized composting of organic waste in a European rural region: A case study in Allariz (Galicia, Spain). In: Mihai F-C, editor. Solid Waste Management in Rural Areas. Rijeka, Croatia: IntechOpen; 2017. pp. 53-79. DOI: 10.5772/intechopen.69555

[8] European Commission. Rural Development in the European Union—Statistical and Economic Information; 2013. pp. 1-390. Available from: https://ec.europa.eu/agriculture/statistics/rural-development/2013_en

[9] Instituto Nacional de Estadistica (Spanish Statistical Office) [Internet]. Available from: https://www.ine.es/

[10] Martínez-Blanco J, Colón J, Gabarrell X, Font X, Sánchez A, Artola A, et al. The use of life cycle assessment for the comparison of biowaste composting at home and full scale. Waste Management. 2010;**30**(6):983-994. DOI: 10.1016/j.wasman.2010.02.023

[11] Zotos G, Karagiannidis A, Zampetoglou S, Malamakis A, Antonopoulos I-S, Kontogianni S, et al. Developing a holistic strategy for integrated waste management within municipal planning: Challenges, policies, solutions and perspectives for Hellenic municipalities in the zero-waste, low-cost direction. Waste Management. 2009;**29**(5):1686-1692. DOI: 10.1016/j.wasman.2008.11.016

[12] Tadesse T, Ruijs A, Hagos F. Household waste disposal in Mekelle city, Northern Ethiopia. Waste Management. 2008;**28**(10):2003-2012. DOI: 10.1016/j.wasman.2007.08.015

[13] Adhikari BK, Trémier A, Martinez J, Barrington S. Home and community composting for on-site treatment of urban organic waste: Perspective for Europe and Canada. Waste Management & Research. 2010;**28**(11):1039-1053. DOI: 10.1177/0734242X10373801

[14] Mihai F-C, Taherzadeh MJ. Introductory chapter: Rural waste management issues at global level. In: Mihai F-C, editor. Solid Waste Management in Rural Areas. Rijeka, Croatia: IntechOpen; 2017. pp. 1-10. DOI: 10.5772/intechopen.70268

[15] Smith SR, Jasim S. Small-scale home composting of biodegradable household waste: Overview of key results from a 3-year research programme in West London. Waste Management & Research. 2009;**27**(10):941-950. DOI: 10.1177/0734242X09103828

[16] European CommissionWorking Document: Biological Treatment of Biowaste, 2nd draft. Directorate-General Environment; 2001. p. 22

[17] Storino F, Menéndez S, Muro J, Aparicio-Tejo PM, Irigoyen I. Effect of feeding Regime on composting in bins. Compost Science & Utilization. 2017;**25**(2):71-81. DOI: 10.1080/1065657X.2016.1202794

[18] Hargreaves JC, Adl MS, Warman PR. A review of the use of composted municipal solid waste in agriculture. Agriculture, Ecosystems & Environment. 2008;**123**(1):1-14. DOI: 10.1016/j.agee.2007.07.004

[19] Zucconi F, de Bertoldi M. Compost specifications for the production and characterization of compost from municipal solid waste. In: de Bertoldi M, Ferranti M, L'Hermite M, Zucconi F, editors. Compost: Production, Quality and Use. London: Elsevier Applied Science; 1987. pp. 30-50

[20] Boletín Oficial del Estado. Real Decreto 506/2013, de 28 de junio, sobre productos fertilizantes (Royal Decree on fertilizers); 15 January 2018; núm. 164, Sect. 1; 2013. pp. 51119-51207

[21] Smith S. A critical review of the bioavailability and impacts of heavy metals in municipal solid waste composts compared to sewage sludge. Environment International. 2009;**35**(1):142-156

[22] Ansorena J. El compost de biorresiduos. Normativa, calidad y aplicaciones. Madrid: Mundiprensa; 2016. 264 p

Chapter 3

The Use of Composted Municipal Solid Waste under the Concept of Circular Economy and as a Source of Plant Nutrients and Pollutants

María Belén Almendro-Candel, Jose Navarro-Pedreño, Ignacio Gómez Lucas, Antonis A. Zorpas, Irene Voukkali and Pantelitsa Loizia

Abstract

The European Union (EU) is one of the major producers of municipal solid wastes and has a common policy based on circular economy to reuse the wastes. However, there are differences between countries and the methods for disposal and treatments. Municipal solid waste (MSW) can be composted and recycled as a source of plant nutrients and improves soil properties. This chapter analyzed the production in the EU and the effects on plant nutrients and environmental pollutants when MSW is added to the soil. The origin of the waste and the compost-like output (CLO) derived is important to determine the expectative of nutrient availability and other possible risks. MSW is so heterogeneous, but after a good pretreatment, an organic-rich matter mix can be composted giving a stabilized organic matter. The addition of the CLO to the soils can improve the nutrient status and favor the bioavailability of nutrients (macronutrients and micronutrients). In general, an increment of N and P was found in the soils. Moreover, important micronutrient availability (Fe, Mn, Cu, and Zn) has been described. However, the presence of pollutants and their mobility should be considered as an environmental risk.

Keywords: circular economy, MSW compost, nitrogen, plant nutrients, pollutants

1. Introduction

In Europe, each of the half billion citizens (500 million people) produces waste. The quantities of municipal solid waste (MSW) have been growing for many years in many countries. This is on top of massive amounts of waste generated from several activities like manufacturing (360 million t) and construction (900 million t), while water supply and energy production generate more than 95 million t [1]. More or less the entire EU produces up to 3 billion t/y in 2011 according to Eurostat [2]. As a definition of MSW [3], "Municipal waste is mainly produced by households, though similar wastes from sources such as commerce, offices and public institutions are included. The amount of municipal waste generated consists of waste collected by or on behalf of municipal authorities and disposed of through the waste management system."

The amount of waste we are creating is increasing, and the nature of waste itself is changing, partly due to the dramatic rise in the use of hi-tech products. According to the latest official Eurostat statistics [2], the total waste generation in the EU-27 was more than 2.62 billion t. The statistics indicated that the total amount of municipal solid waste is continuously rising [4–6] and the amount up to 98 million t (or 3.7%) was classified as hazardous. On 2008, each European citizen produced more or less 5.2 t/y of waste, of which 196 kg were hazardous [2]. As indicated from the Organization for Economic Co-operation and Development (OECD) [7], MSW increased up to 54% in major EU countries such as Switzerland, Denmark, Portugal, the Netherlands, and Greece in 20 years (1980–2000). OECD [7], Jacobsen and Kristoffersen [8], and Zorpas et al. [9] investigated the connection between economic growth and quantity of waste and proposed that a decoupling is needed in order to reduce the increasing burden from waste management.

MSW from 2000 has slightly minimized in the EU-27, although the gross domestic product (GDP) was increased by 33% between 2000 and 2013, due to economic crisis [9]. However, waste generation in new member states has remained relatively stable by weight since the 1990s. This may be due to a reduced incidence of heavy mining and construction waste and increased lighter paper and packaging waste. Decoupling economic growth from the environmental impacts associated with waste generation is a key objective of the EU [10]. The target is not only to monitor the generation of waste but also to reduce the waste production [5].

1.1 Production of MSW in the EU countries

Among the EU countries, there are huge differences in the production as well as in the treatment of MSW. The average production per country varies from 254 kg/y in Romania to 758 kg/y in Denmark with the average to be 474 kg/y. Cyprus produced approximately 630 kg/y, Greece 650 kg/y, and Spain 495 kg/y (**Figure 1**).

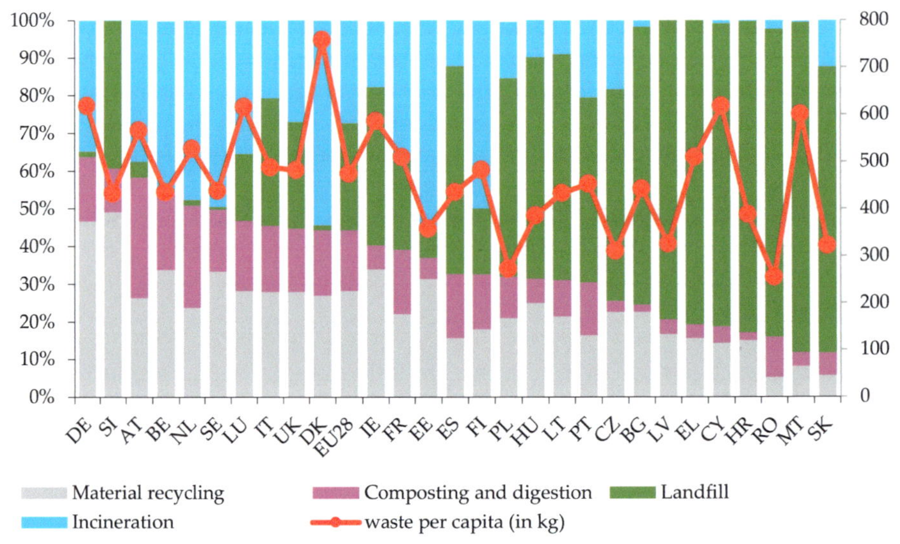

Figure 1.
Waste production and management in the EU countries.

Municipal waste per capita in the EU decreased from 523 kg per person in 2007 to 474 kg per person in 2014, in part because of the economic downturn.

The share of recycled or composted municipal waste in the EU-28 (including Croatia) increased from 31% in 2004 to 44% in 2014. According to the European Environment Agency (EEA) [3], trends in the past decade also include a shift away from landfilling and a 56% drop in net greenhouse gas emissions from municipal waste management between 2001 and 2010. Recycling and composting range from 64% in Germany to 12% in Slovakia and Malta (EU average, 44%). Six member states landfill less than 5% of their municipal waste, 8 member states landfill over 70% of their municipal waste (EU average, 28%), 10 member states incinerate over 35% of their municipal waste, and 8 member states incinerate less than 2% of their municipal waste (EU average, 27%) [11]. The overall increase in the recycling rate appears in some items like paper/cardboard, glass, metals, plastics, and textiles. In contrast, increases in biowaste recycling are much more modest [3].

Packaging waste in the EU in 2011, measured by weight, is made up of paper and cardboard (40%), glass (20%), plastic (19%), wood (15%), and metal (6%), according to Eurostat [2]. In 2013, 65% of packaging was recycled in the EU-28, although material-specific recycling rates varied a great deal: 85% for paper and cardboard packaging, 74% for metallic packaging, 73% for glass packaging, 36% for wooden packaging, and 37% for plastic packaging. Moreover, in yearly base almost 9 million t of end-of-life vehicles (ELV) are generated in the EU and can be recovered almost 80% of ELV materials [11, 12].

A significant issue of MSW is the food waste (FW), and according to FAO [13], in 2011 it is estimated that 35% of food (including supply chain) is mostly lost at the consumer level. Moreover, 1.3 billion t of edible foodstuffs (equivalent with one-third of the global food production) are lost yearly [13, 14], and this is sufficient to feat one-eighth of worldwide population [15]. Additionally, the total CO_2 equivalences of greenhouse gases (GHG) from the entire FW is about 3.49 billion t [15], and the annual bulk-trade value of produced and unconsumed food is estimated at 936 billion $.

The management of MSW is an increasing problem in small communities as well as in insular communities such as (Malta, Crete, Sicily, and Cyprus) because of the fast increase in population density, which is leading to the collapse of landfill sites [9]. It is open of question nowadays as indicated by Zorpas et al. [15], "how a small island will implement the concept of circular economy" with all the ambitious targets that were set. This perspective presents a significant challenge for any insular community as the European Union Landfill Directive has presented stringent requirements for waste disposal sites and requires a reduction for waste (biodegradable) being dumped [16].

1.2 Circular economy and wastes

According to Winans et al. [17], there are limited data about the clear evidence of the origin of the concept of circular economy. However, according to Ellen MacArthur Foundation [18], some contributions include researches from the United States as may also have been stimulated by Rachel Carson's Silent Spring [19] , which states that "limits to growth" thesis of the Club of Rome in the 1970s, the "spaceship earth" metaphor presented by Barbara Ward and Kenneth Boulding, and work by eco-economist Herman Daly [20]. Pearce and Turner [21] proposed the general framework of circular economy with emphasis on product resource and pollution. The main principles were presented by Zorpas and Lasaridi [5], Wu et al. [22], and Zorpas et al. [23], and more specifically the well-known 3Rs (reduce, reuse, recycle) and the 6Rs (reuse, recycle, redesign, remanufacture, reduce, recover) by Jawahir and Bradley [24]. Moreover, Zorpas [25] indicated the concept of "11R," which starts from refuse and ends to recover.

Waste generation is the other side of the coin of resource exploitation and potential scarcity. Therefore, it is interwoven with global environmental security and governance, posing a problem that has grave environmental, social, and economic repercussions for all nations, for the current and future generations.

The concept of circular economy appeared in Europe in 1980 and 1990 with several other policies that also appear in the EU drawing on ideas that can be traced to 1970 [26]. Following the concern around high commodity prices, the European Commission (EC) launched a *flagship* initiative on resource efficiency, which at the beginning was operationalized through the roadmap for a resource-efficient Europe [27]. This was followed up with the declaration of a range of policy measures known cooperatively as the Circular Economy Package.

During 2014, the European Commission (EC) published a statement entitled "Toward a circular economy: A zero waste program for Europe." This report provides emphasis on the "the EU and the Member States should encourage investment in circular economy innovation and its take-up" [28]. Nevertheless, before the end of 2014, the proposals on the circular economy were eliminated as part of the drive to cut red tape [29]. During 2015 a new proposal focused on circular economy was realized by the European Commission. The new proposal entitled "Circular Economy closing the Loop – An EU Action Plan for the Circular Economy" sets out the new targets, the policies on the circular economy [30].

The EC's action plan for the circular economy has an ambitious goal: "to treat waste as a resource and to turn Europe into a circular economy." Although the recommended policies go far beyond the waste division, waste division management plays a key role in the transition to a circular economy. As such, the EC's 2015 action for a circular economy sets the current scene for a new approach to waste management in Europe.

The action plan sets out a policy framework that builds on and integrates existing policies and legal instruments. In particular, the European Circular Economy Action Plan proposes amendments to legislation relating to waste and landfills (which were due for revision). Changes on the following legislations were proposed by the EC in order to turn Europe into a circular economy: (i) Directive 1999/31/EC [16] on the landfill of waste, (ii) Waste Framework Directive (WFD) 2008/98/EC [31] on waste, (iii) Directives 2000/53/EC [32] on end-of-life vehicles, 2006/66/EC [33] on batteries and accumulators and waste batteries and accumulators, 2012/19/EU [34] on waste electrical and electronic equipment (WEEE), and (iv) Directive 94/62/EC [35] on packaging and packaging waste. The action plan suggests three specific changes to the regulations by including the following targets by 2030: (a) a target to prepare 65% of municipal waste for reuse and recycling, (b) a binding landfill target to reduce landfill to a maximum of 10% of municipal waste, and (c) a target to prepare 75% of packaging waste for reuse and recycling by 2030 (with supplementary targets for specific packaging material).

On the one hand, it is important to achieve a reuse or recycling of 65% of MSW and reduce binding landfill. On the other hand, the use of MSW in soils as a source of nutrients and the main way to reuse the organic matter is not an optional target in the EU and for extension to other countries (**Figure 2**). In fact, this is an essential part of the circular economy, and the role of the administrations to ensure this use is crucial. However, we should consider that the "requirements that have to be content by a material derived from waste to confirm that the quality of the material is such that its use is not detrimental for human health or the environment" [36].

Considering developing and promoting recycling in the concept of circular economy, the main fraction of MSW is organics; those could be very useful to

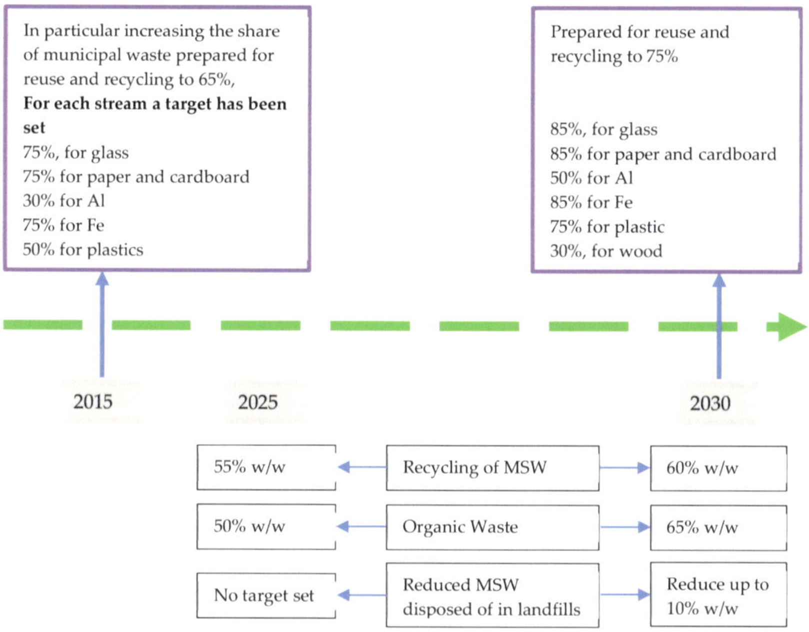

Figure 2.
Circular economy time line from 2015 to 2030.

provide several nutrition to soils as an acceptable method to treat them is composting [37, 38]. However, composts should cease to be waste only if they are placed on the market for specific purpose and only if acceptable criteria will be given [6]. Creating compost delivers economic and more specific ecological and environmental benefits. The production of compost with high and reliable quality expands its use and avoids unnecessary regulatory burden or other legal certainties. Nowadays, the quality of composted materials is determined only by the end use and classified according to its physicochemical characteristics. Having in mind the cure of circular economy and industrial symbiosis, the development of end waste criteria (EWC) for any organic material before the production of compost could be extremely helpful. For example, a bad quality compost with low C/N ratio or low organic matter or low bulking density could be useful for restoration of mining activities [39].

2. Macronutrients and environmental pollution

The application of organic amendments to the soil is a very common practice, especially in areas with low organic matter content [39]. The application of MSW as a source of organic matter and nutrients has been described for agriculture, mining restoration, and gardening [39, 40]. But it carries the associated risk of possible pollution, focused mainly in the nitrate contamination of surface and groundwater, since the mineralization of this organic matter can release large amounts of ammonium that will oxidize to nitrate [41, 42]. However, there are also other risks derived from the composition of MSW (hazardous materials) and the presence of plant nutrients as phosphorus [43], chloride, and sulfur [44].

A common composition of a composted MSW is indicated in **Table 1**. This composition shows important amounts of plant nutrients (i.e., phosphorus) as well as the presence of environmental pollutants like nickel and cadmium [39].

The urban wastes can differ in origin and changes due to the different style of life, conditioning the composition, and the total amount of wastes. The

Variables	Amounts[a]
Sand (20 < Ø < 2000 μm)	42%
Silt (2 < Ø < 20 μm)	28%
Clay (<2 μm)	30%
pH in water (1:2.5)	6.9
Electrical conductivity (EC) (1:5)	705 dS/m
Oxidizable organic matter (OM)	416 g/kg
Phosphorus (P)	4610 mg/kg
Potassium (K)	2100 mg/kg
Sodium (Na)	1010 mg/kg
Calcium (Ca)	60 mg/kg
Magnesium (Mg)	45 mg/kg
Iron (Fe)	9800 mg/kg
Manganese (Mn)	177 mg/kg
Copper (Cu)	89 mg/kg
Zinc (Zn)	186 mg/kg
Nickel (Ni)	18.8 mg/kg
Cadmium (Cd)	0.8 mg/kg

[a]*Amounts on a dry matter basis.*

Table 1.
Composition of a composted MSW [39].

composition of waste in landfills could differ due to the joint storage of industrial and domestic waste containing toxic elements [45]. Moreover, the composition of MSW can be different considering the seasons of the year and seasonal impacts should be taken into consideration when dealing with MSW [46].

2.1 Nitrogen and organic matter

Nitrogen is one of the major nutrients for plants, and soil is the main source in terrestrial ecosystems. Nitrate is the preferable chemical form for the absorption of most of the plants. However, this is a very mobile chemical form [42]. In order to minimize the risk of groundwater contamination, Jorge-Mardomingo et al. [41] recommend the use of stable organic amendments (with a more stabilized organic matter), which could produce a lower content of leachable nitrogen forms. Risk is also minimized by planting rainfed crops and particularly by choosing crops with a high demand for nitrogen such as wheat or maize [47].

Applications of MSW (composted or not) should be planned to avoid the coincidence of peaks of soluble nitrogen forms with rainfall periods in order to prevent their transport to groundwater and increase their residence time in the root zone. Diffuse nitrogen losses from agricultural fields are the major cause of excessive nitrate concentrations in ground- and surface waters [48].

MSW compost contains large amounts of organic matter and both organic nitrogen and inorganic nitrogen [49]. The organic matter plays a key role in improving soil properties such water retention capacity or soil structure, among others [50]. The use of composted organic wastes produces changes in soil physical, chemical, and biological properties and can enhance plant growth after its application [51].

Moreover, the organic matter added with MSW can be the main source of nitrogen in impoverished soils with low organic matter content.

The amounts of plant-available nitrogen and phosphorus from MSW are closely related to the degree of compost maturity, the addition of mineral fertilizers, soil characteristics, and environmental parameters [49]. All of them can affect the availability of nutrients. For instance, the addition of inorganic fertilizers can increase the plant and microbial activity of soils and may induce an increment of the mineralization of the organic matter of MSW, favoring the inorganic nitrogen forms.

Not only the plant nutrition is directly affected by using MSW, but also improving soil properties, the plant can response positively. Civeira [51] studied the response of an urban-degraded soil to different MSW compost application rates, as an alternative to MSW disposal and soil recovery. As indicators from soil response, physical (bulk density, soil moisture, and water infiltration) and chemical (pH, electrical conductivity, organic C, total N, and extractable P) parameters were evaluated. Compost application positively affected total N content in soils, improving soil physical properties in a similar way to chemicals, after MSW compost addition.

After the application of MSW compost to the soil, nitrogen is transformed into mobile forms, which can be accumulated in the soil, absorbed by plants, or released into the atmosphere or water system. The amount of nitrogen released into the soil solution determines the form of nitrogen availability to the plant and, consequently, the yield. Nevertheless, the environmental risks are well known. The amendment of the soil with organic fertilizers containing easily decomposable organic carbon compounds can trigger denitrification processes [50].

If MSW is poor in nitrogen or the rate C/N is inadequate for the mineralization of the organic matter, additional sources of nitrogen are needed. Mkhabela and Warman [52] found that the low availability of compost-N means that supplementary nitrogen in the form of inorganic fertilizer may have to be added together with compost in order to enhance N availability to crops. They observed that inorganic fertilizer (NPK) and a mixture of MSW compost and inorganic fertilizer produce higher yields than MSW compost alone.

2.2 Phosphorus and other macronutrients

Some authors observed that MSW compost effectively supplies phosphorus to soil with its concentration increased when increasing application rates. MSW composts provided equivalent amounts of phosphorus to soil as mineral fertilizers [52, 53].

In an experiment in plots in a quarry restoration, where 3 kg/m^3 of MSW were applied to a substrate composed by limestone outcrop from the rejection of the quarry, an increment of nutrients associated to the composition of the composted MSW was obtained. In the plots in which MSW was applied, an important increase in the soil content of N-Kjeldahl, available P, and the rest of macro- and micronutrients was found, favoring the plant growth [54]. The results reflected the contribution of MSW to the plant nutrition and reinforced the idea of the positive use of the organic fraction of MSW in mining and landfill restoration (**Figure 3**).

Baldi et al. [55] studied the effect of applying 5 and 10 t dw/ha·year of composted MSW to a nectarine crop for 11 years. They found that the content of N, P, macro-, and micronutrients increased with respect to the control, both in the plant and in the fruit. The authors concluded that in their experiment the slow release of nutrients in the soil from compost mineralization seemed to match with plant demand, supporting the hypothesis that compost can be used effectively in fruit

Figure 3.
Composted MSW for soil restoration in a landfill area and a quarry with compost derived from MSW. The use of compost of MSW for seed germination (Photos from J. Navarro Pedreño).

tree nutrient management, since it promotes an increase of tree growth and yield by maintaining an optimal nutritional status of plants.

Calleja-Cervantes et al. [56] studied the effect that 13 years of applying three different composted organic amendments have had on soil quality, GHG emissions, and the dynamics of its microbial communities 15 days after the annual application. They found that total nitrogen increased with respect to the control by amending with organic fraction of municipal solid waste. Organic amendment application resulted in higher levels of phosphorus and potassium in the soil. They concluded that significantly higher organic matter contents, total N, P, and K contents, in the soil when compared to the control validate the fact that organic waste-based fertilizers contribute to enhanced soil fertility.

The balance between the addition of nutrients that can be available for plant nutrition and the possible pollution, especially of waters with N-forms, needs to study previously the type of soil and, in general, the environmental conditions where MSW is going to be applied. The criteria established to control the addition of MSW as amendment to the soil might be improved including new criteria based on environmental conditions.

3. Micronutrients and trace elements from MSW

As it has been shown in the previous sections, the use of MSW can be very positive due to the addition of plant essential elements and the availability of them in the soils and due to the improvement of some physical properties [50]. However, trace elements should be identified and considered as environmental risk.

In the EU, as in the rest of the world, several treatments are used for MSW, mainly landfill disposal. However, landfill and composting are not the only treatments for urban wastes. Incineration has been increased (with or without recovering energy), and it is an important treatment used with MSW in the EU.

It is important to consider that countries with limited natural resources should have an interest in resource reuse [57] and the addition of MSW to soil has positive benefits. For these reasons, the composting process for municipal solid waste should be implemented as far as possible due to the great organic fraction of MSW.

Composts have been frequently used as nitrogen and organic carbon amendments to improve soil quality and to support plant growth, with the additional benefit of reducing waste disposal costs [49]. Nevertheless, the environmental risks from the use of MSW begin within the previous treatments before its addition to soil. The composting process is recommendable before its use, although health risk assessment of odor emissions (i.e., sulfides and aromatics) from waste composting is important [58].

Regarding with the major potential environmental impacts related to landfill, the main problem identified in the municipal wastes consisted of untreated leachates [59]. The leachate pollution of groundwater and surface waters can be categorized into four groups (dissolved organic matter, inorganic macrocomponents, heavy metals, and xenobiotic organic compounds) [60]. Kjeldsen et al. [60] defined these groups for MSW landfill leachates as follows:

- Dissolved organic matter, quantified as chemical oxygen demand (COD) or total organic carbon (TOC), volatile fatty acids, and more refractory compounds such as fulvic-like and humic-like compounds.

- Inorganic macrocomponents: calcium (Ca^{2+}), magnesium (Mg^{2+}), sodium (Na^{+}), potassium (K^{+}), ammonium (NH_4^{+}), iron (Fe^{2+}), manganese (Mn^{2+}), chloride (Cl^{-}), sulfate (SO_4^{2-}), and hydrogen carbonate (HCO_3^{-}).

- Heavy metals: cadmium (Cd^{2+}), chromium (Cr^{3+}), copper (Cu^{2+}), lead (Pb^{2+}), nickel (Ni^{2+}), and zinc (Zn^{2+}).

- Xenobiotic organic compounds (XOCs) originating from household or industrial chemicals and present in relatively low concentrations (usually less than 1 mg/l of individual compounds). These compounds include among others a variety of aromatic hydrocarbons, phenols, chlorinated aliphatics, pesticides, and plasticizers.

The inorganic macrocomponents can be complemented considering other nitrogen forms derived from the oxidation of ammonium that are easily leachate as nitrate (NO_3^{-}) and nitrite (NO_2^{-}) due to the composition of MSW (organic matter is the major fraction) and its biodegradation. The overuse of nitrogen fertilizer can cause the leaching of NO_3^{-} to the surrounding water source and the emissions of N_2O and NO to the atmosphere [61].

The use of MSW as soil amendment, after a good composting process, can produce several environmental risks, which can be summarized as:

- The excess of nutrients/pollutants/organic-soluble compounds that can affect waters and plants

- The persistence of undesirable objects and fragments of objects, plastics, glass, and other materials that are difficult to biodegrade in the soil

Moreover, the presence of fragments and objects in MSW is a major concern related to the use of MSW as soil amendment because of the addition to the topsoil of undesirable objects. Farmers and other potential users (i.e., gardeners) do not want to use amendments and fertilizers that look unpleasant and contain materials that cannot be easily integrated into the soil.

Most of these problems (presence of solid fragments of major size, XOCs, etc.) can be solved in the treatments previously carried out on urban waste treatment plants, including sieving processes before and after composting. The problem of MSW is more serious in developing countries [62] without the application of an adequate treatment to the municipal wastes before applying them to soils.

The excess of nutrients and other elements like trace pollutants is a more difficult problem to solve in the treatments carried out in the municipal waste plants before the application of MSW. The urban waste is usually composted before soil addition facilitating the stabilization of the organic compounds although its application has environmental risks due to soluble organic carbon forms [63], nutrients, and the increment of pollutants, especially by leaching them to waters. Yusof et al. [63] found direct influences of leachate from MSW in the form of inorganic nitrogen and heavy metals in waters.

In general, after an adequate treatment of MSW, we pay our attention in the pollution of the soil–plant system and water, due to the excess of micronutrient or pollutants available from this waste.

The heavy metal pollution of surface soil horizons is characteristic for the sites of solid waste storage and their impact zones irrespectively of climatic conditions, ways of waste management, and stages of the life cycle [45]. At the same time, heavy metals accumulate in ruderal herbaceous plants [45]. However, soil moisture, irrigation, and climate conditions (rain) can affect the mobility and displacement of pollutants to surface water and groundwater and favor their presence in the root environment. In this case, it is possible to incorporate the pollutants into the food chain by plant uptake.

So, there would be a serious risk associated with the availability and mobility of trace elements, including the excess of micronutrients. In general, the addition of MSW increases the presence of trace elements in the soil [62]. Long-term application of municipal solid waste compost may result in accumulation of toxic metals in amended soil, as it has been demonstrated [64].

In general, an increment in leaching and changes in plant composition have been observed and can lead to environmental problems related to water contamination and the accumulation in the food chain of trace elements. Rezapour et al. [62] observed that soils were significantly enriched by the available and total fractions of the metals in the sequences of Zn > Pb > Ni > Cd > Cu and Cd > Zn > Ni > Pb > Cu, respectively. Nevertheless, only the Cd content exceeded the standard levels. However, many works found Cd-soluble concentrations in leachate below the detectable rates and an increment of the soluble fractions of Zn and Ni [65]. Cu and Zn availability is increased with MSW [66], as well as it has been reported for landfill sites [67].

Trace elements are accumulated in different parts of the plants. For instance, Cd, Cr, and Pb were accumulated in roots and stems in mulberry trees [64] and the Cu and Zn concentration in grains of wheat [68]. Adamcová et al. [69] found the highest degree of accumulation for Cd under the use of MSW. Cd, as well as Cr, Ni and Zn are accumulated mostly in the leaves, whereas Co, Cu, Fe, Hg, Mn, and Pb are accumulated mostly in the roots in the case of tansy (*Tanacetum vulgare* L.).

The application of MSW and derived materials from them, as the compost-like output (CLO) is most of the times used based on the nitrogen content as it is an important parameter for soil fertilization. However, metal pollution should be considered as heavy metal concentrations could exceed water quality limits at the higher application rates. This was found when applying amounts over 3000 kg N/ha [70]. However, the type of soil and the irrigation are important factors that can control the pollution to waters.

Another negatives effects were described. Leachates also pose pressures on biochemical and chemical oxygen demand (BOD and COD), TOC, ammonium and sulfur compositions, and heavy metals in soil and groundwater [71].

Salinity of soils and water can be increased by using MSW and biosolids [44, 69]. Hamidpour et al. [68] detected after a 3-year experiment the increment of soil salinity due to the use of MSW. This means that soluble salts (inorganic ions) are presented in the soil solution derived from MSW. However, in saline soils, MSW compost, with high organic matter content and low concentrations of inorganic and organic pollutants, allows an improvement of physical, chemical, and biochemical characteristics and constitutes low-cost soil recovery [72].

Biological activity of soils can be affected by the addition of MSW, both in positive and negative ways. Farrell et al. [73] showed the increment of microbial activity in contaminated soils with Cu, Pb, and Zn. Composts can successfully immobilize heavy metals and promote ecosystem diversity/function; surface incorporation had little remedial effect below the surface layer over the course of our short-term trial. On the other hand, the presence of XOCs can alter the soil biota.

It is obvious that there are environmental risks associated to the use of MSW due to the possible pollution of water, the plant uptake of pollutants with an impact in the food chain, and the presence of undesirable fragments. Nevertheless, the use of MSW compost, considering these risks and the type of soil where it is applied, can be controlled or minimized the risks.

4. Conclusions

MSW is a worldwide problem, even if we are able to reduce the amount produced every year. However, developing countries are increasing the MSW production parallel to the effort of developed countries to reduce their production. In the EU countries, there is no harmonization of the treatments applied in each country, but all of them are promoting the reuse based on the circular economy. In the case of MSW, the composting process of the important organic fraction is one of the best strategies to improve soil properties, reduce the landfill disposal of this waste, and recycle the nutrients. However, two important environmental risks should be considered even if MSW is well pretreated and composted. The excess of nutrients can contaminate waters and the presence of trace elements can has a negative effect on the food chain.

The risks can be minimized if pre- and posttreatments on MSW are applied and if the composted matter is added in a soil under adequate environmental conditions. This is a key factor to improve circular economy and ensure the use of these wastes. Nitrogen and other nutrients, especially micronutrients, are presented in MSW and can be bioavailable for plants. Nevertheless, in the same way, pollutants with emphasis in trace elements increase their availability.

The criteria for the recycling of MSW and compost-like output (CLO) derived as amendment should consider the end use but the previous treatments of MSW and have to look for a balance between the input of nutrients and the environmental risks associated with the soil conditions.

Author details

María Belén Almendro-Candel[1]*, Jose Navarro-Pedreño[1], Ignacio Gómez Lucas[1], Antonis A. Zorpas[2], Irene Voukkali[2] and Pantelitsa Loizia[2]

1 Department of Agrochemistry and Environment, Miguel Hernández University of Elche, Elche (Alicante), Spain

2 Faculty of Pure and Applied Sciences, Environmental Conservation and Management, Laboratory of Chemical Engineering and Engineering Sustainability, Cyprus Open University, Nicosia, Cyprus

*Address all correspondence to: mb.almendro@umh.es

References

[1] EC (European Commission). Being wise with waste: The EU's approach to waste management. 2010. Available from: http://ec.europa.eu/environment/waste/pdf/WASTE%20BROCHURE.pdf [Accessed: September 24, 2018]

[2] Eurostat [Internet]. 2011. Available from: http://epp.eurostat.ec.europa.eu/statistics_explained/index.php/Waste_statistics [Accessed: May 12, 2015]

[3] EEA (European Environment Agency). Managing municipal solid waste: A review of achievements in 32 European Countries. 2013. Available from: https://www.eea.europa.eu/publications/managing-municipal-solid-waste [Accessed: October 02, 2018]

[4] Salhofer S, Obersteiner G, Schneider F, Lebersorger S. Potentials for the prevention of municipal solid waste. Waste Management. 2008;**28**:245-259. DOI: 10.1016/j.wasman.2007.02.026

[5] Zorpas AA, Lasaridi K. Measuring waste prevention. Waste Management. 2013;**33**:1047-1056. DOI: 10.1016/j.wasman.2012.12.017

[6] Zorpas AA, Lasaridi K, Voukkali I, Loizia P, Chroni C. Household waste compositional analysis variation from insular communities in the framework of waste prevention strategy plans. Waste Management. 2015;**38**:3-11. DOI: 10.1016/j.wasman.2015.01.030

[7] OECD. OECD workshop on waste prevention: toward performance indicators (No. ENV/EPOC/WGWPR/SE(2002)1/FINAL). In: Proceedings of the OECD Workshop in Waste Prevention: Toward Performance Indicators; 8-10 October 2001; Paris (France). Paris: OECD; 2002. 237 p. Available from: http://www.oecd.org/officialdocuments/publicdisplaydocumentpdf/?doclanguage=en&cote=env/epoc/wgwpr/se(2002)1/FINAL [Accessed: September 25, 2018]

[8] Jacobsen H, Kristoffersen M. Case Studies on Waste Minimisation Practices in Europe. Copenhagen: European Environment Agency; 2002. 52 p. Available from: https://www.eea.europa.eu/publications/topic_report_2002_2 [Accessed: September 20, 2018]

[9] Zorpas AA, Lasaridi K, Abeliotis K, Voukkali I, Loizia P, Fitiri L, et al. Waste prevention campaign regarding the waste framework directive. Fresenius Environmental Bulletin. 2014;**23**(11a):2876-2883

[10] Mazzanti M, Zoboli R. Waste generation, waste disposal and policy effectiveness: Evidence on decoupling from the European Union. Resources, Conservation and Recycling. 2008;**52**:1221-1234. DOI: 10.1016/j.resconrec.2008.07.003

[11] Bourguignon D. Briefing EU Legislation in progress: Circular Economy package. Four legislative proposals on waste. 2016. Available from: http://www.europarl.europa.eu/EPRS/EPRS-Briefing-573936-Circular-economy-package-FINAL.pdf [Accessed: September 14, 2018]

[12] Zorpas AA, Inglezakis JI. Automotive industry challenges in meeting EU 2015 environmental standard. Technology in Society. 2012;**34**:55-83. DOI: 10.1016/j.techsoc.2011.12.006

[13] Gustavsson J, Cederberg C, Sonesson U, van Otterdijk R, Meybeck A. Global Food Losses and Food Waste—Extent Causes and Prevention. Food and Agriculture Organization of the United Nations: Rome; 2011. 37 p. Available from: http://www.fao.org/

docrep/014/mb060e/mb060e00.htm [Accessed: September 10, 2018]

[14] Chalak A, Abou-Daher C, Chaaban J, Abiad GM. The global economic and regulatory determinants of household food waste generation: A cross-country analysis. Waste Management. 2016;**48**:418-422. DOI: 10.1016/j.wasman.2015.11.040

[15] Zorpas AA, Lasaridi K, Pociovălişteanu MD, Loizia P. Household compost monitoring and evaluation in the framework of waste prevention strategy. Journal of Cleaner Production. 2018;**172**:3567-3577. DOI: 10.1016/j.jclepro.2017.03.155

[16] Council of the European Union. Council Directive 1999/31/EC of the 26 April 1999 on the landfill of waste. Official Journal of the European Communities. 1999;**L182**:1-19. Available from: https://eur-lex.europa.eu/legal-content/EN/TXT/PDF/?uri=CELEX:31999L0031&from=en [Accessed: February 15, 2018]

[17] Winans K, Kendall A, Deng H. The history and current applications of the circular economy concept. Renewable and Sustainable Energy Reviews. 2017;**68**:825-833. DOI: 10.1016/j.rser.2016.09.123

[18] Ellen MacArthur Foundation. Toward the Circular Economy, vol. 2: Opportunities for the Consumer Goods Sector [Internet]. 2015. Available from: http://www.ellenmacarthurfoundation.org/business/reports/ce2013 [Accessed: August 05, 2018]

[19] Carson R. Silent Spring. Greenwich, CT: Fawcett Publications; 1962. 400 p

[20] Naustdalslid J. Circular economy in China—The environmental dimension of the harmonious society. International Journal of Sustainable Development and World Ecology. 2014;**21**(4):303-313. DOI: 10.1080/13504509.2014.914599

[21] Pearce DW, Turner RK. Economics of Natural Resources and the Environment. Baltimore, MD: JHU Press; 1990. 392 p

[22] Wu HQ, Shi Y, Xia Q, Zhu WD. Effectiveness of the policy of circular economy in China: A DEA-based analysis for the period of 11th five-year-plan. Resources, Conservation and Recycling. 2014;**83**:163-175. DOI: 10.1016/j.resconrec.2013.10.003

[23] Zorpas AA, Voukkali I, Navarro-Pedreño J. Tourist area metabolism and its potential to change through a proposed strategic plan in the framework of sustainable development. Journal of Cleaner Production. 2018;**172**:3609-3620. DOI: 10.1016/j.jclepro.2017.02.119

[24] Jawahir IS, Bradley R. Technological elements of circular economy and the principles of 6R-based closed-loop material flow in sustainable manufacturing. Procedia CIRP. 2016;**40**:103-108. DOI: 10.1016/j.procir.2016.01.067

[25] Zorpas AA. How the Proposed EU Strategy of Circular Economy contributes to Environmental "Footprint" Boosting Bio Products into Sustainability. In: ECO-BIO 2018 Conference; 4-7 March 2018; Dublin (Ireland). 2018

[26] McDowall W, Ceng Y, Huang B, Bartekova E, Bleischwitz R, Turkeli S, et al. Circular economy policies in China and Europe. Journal of Industrial Ecology. 2017;**21**(3):651-661. DOI: 10.1111/jiec.12597

[27] EC (European Commission). Roadmap to a resource efficient Europe. COM(2011) 571 final. 2011. Available from: https://eur-lex.europa.eu/legal-content/EN/TXT/?uri=CELEX:52011DC0571 [Accessed: Spetmber 07, 2018]

[28] Euractiv. Commission push against red tape scores points in Parliament [Internet]. 2015. Available from: http://www.euractiv.com/section/science-policymaking/news/commission-push-against-red-tape-scores-points-in-parliament/ [Accessed: September 24, 2018]

[29] EC (European Commission). Closing the loop—An EU action plan for the Circular Economy COM (2015) 614 final. 2015. Available from: http://eur-lex.europa.eu/resource.html?uri=cellar:8a8ef5e8-99a0-11e5-b3b7-01aa75ed71a1.0012.02/DOC_1&format=PDF [Accessed: September 24, 2018]

[30] Euractiv. New Circular Economy Package to create fewer jobs than axed bill it replaced. [Internet]. 2015. Available from: http://www.euractiv.com/section/science-policymaking/news/new-circular-economy-package-to-create-fewer-jobs-than-axed-bill-it-replaced/ [Accessed: September 24, 2018]

[31] European Parliament and the Council of the European Union. Directive 2008/98/EC of the European Parliament and of the Council of 19 November 2008 on waste and repealing certain directives. 2008;**L312**:3-30. Available from: https://eur-lex.europa.eu/legal-content/EN/TXT/PDF/?uri=CELEX:32008L0098&from=EN [Accessed: February 15, 2018]

[32] European Parliament and the Council of the European Union. Directive 2000/53/EC of the European Parliament and of the Council of 18 September 2000 on end-of-life vehicles. 2000;**L269**:34-42. Available from: https://eur-lex.europa.eu/resource.html?uri=cellar:02fa83cf-bf28-4afc-8f9f-eb201bd61813.0005.02/DOC_1&format=PDF [Accessed: March 22, 2008]

[33] European Parliament and the Council of the European Union. Directive 2006/66/EC of the European Parliament and of the Council of 6 September 2006 on batteries and accumulators and waste batteries and accumulators and repealing directive 91/157/EEC. 2006;**L266**:1-14. Available from: https://eur-lex.europa.eu/legal-content/EN/TXT/PDF/?uri=CELEX:32006L0066&from=EN [Accessed: March 15, 2018]

[34] European Parliament and the Council of the European Union. Directive 2012/19/EU of the European Parliament and of the Council of 4 July 2012 on waste electrical and electronic equipment (WEEE). 2012:**L197**:38-71. Available from: https://eur-lex.europa.eu/legal-content/EN/TXT/PDF/?uri=CELEX:32012L0019&from=EN [Accessed: March 15, 2018]

[35] European Parliament and the Council of the European Union. European Parliament and Council Directive 94/62/EC of 20 December 1994 on packaging and packaging waste. 1994;**L365**:10-23. Available from: https://eur-lex.europa.eu/legal-content/EN/TXT/PDF/?uri=CELEX:31994L0062&from=EN [Accessed: March 27, 2018]

[36] Muchová L, Eder P. JRC Scientific and Technical Reports: End-of-Waste Criteria for Aluminum and Aluminum Alloy Scrap: Technical Proposals. Luxembourg: Publications Office of the European Union; 2010. 69 p. Available from: http://publications.jrc.ec.europa.eu/repository/bitstream/JRC58527/jrc58527.pdf [Accessed: March 18, 2018]

[37] Zorpas AA, Kapetanios E, Vlyssides A, Zorpas AG, Karlis KP, Haralambous I, et al. Compost produced from organic fraction of municipal solid waste, primary stabilized sewage sludge and natural zeolite. Journal of Hazardous Materials. 2000;**B77**:149-159. DOI: 10.1016/S0304-3894(00)00233-8

[38] Zorpas AA, Loizidou M. Sawdust and natural zeolite as a bulking agent for the improving the quality of a co-composted product which it based on anaerobically stabilized sewage sludge. Bioresource Technology. 2008;**99**(16):7545-7552. DOI: 10.1016/j.biortech.2008.02.014

[39] Almendro-Candel MB, Navarro-Pedreño J, Jordan MM, Gómez I, Meléndez-Pastor I. Use of municipal solid waste compost to reclaim limestone quarries mine spoils as soil amendments: Effects on Cd and Ni. Journal of Geochemical Exploration. 2014;**144**:263-366. DOI: 10.1016/j.gexplo.2013.11.00

[40] Ozcan HK, Guvenc SY, Guvenc L, Demir G. Municipal solid waste characterization according to different income levels: A case study. Sustainability. 2016;**8**:1044. DOI: 10.3390/su8101044

[41] Jorge-Mardomingo I, Jiménez-Hernández ME, Moreno L, de la Losa A, de la Cruz MT, Casermeiro MA. Application of high doses of organic amendments in a Mediterranean agricultural soil: An approach for assessing the risk of groundwater contamination. Catena. 2015;**131**:74-83. DOI: 10.1016/j.catena.2015.03.013

[42] Pérez-Gimeno A, Navarro-Pedreño J, Almendro-Candel MB, Gómez I, Jordán MM. Nitrogen compound pollution associated with the use of sewage sludge compost and limestone outcrop residue under saline irrigation. Journal of Soils and Sediments. 2016;**16**:1345-1351. DOI: 10.1007/s11368-014-0972-7

[43] Sokka L, Antikainen R, Kauppi P. Flows of nitrogen and phosphorus in municipal waste: A substance flow analysis in Finland. Progress in Industrial Ecology. 2004;**1**(1-3):165-186. DOI: 10.1504/PIE.2004.004677

[44] Pérez-Gimeno A, Navarro-Pedreño J, Almendro-Candel MB, Gómez I, Jordán MM. Environmental consequences of the use of sewage sludge compost and limestone outcrop residue for soil restoration: Salinity and trace elements pollution. Journal of Soils and Sediments. 2016;**16**:1012-1021. DOI: 10.1007/s11368-015-1288-y

[45] Zamotaev IV, Ivanov IV, Mikheev PV, Belobrov VP. Assessment of the state of soils and vegetation in areas of landfills and municipal solid waste sites (a review). Eurasian Soil Science. 2018;**51**(7):827-842. DOI: 10.1134/S1064229318070104

[46] Zhang HB, Zhang HY, Wang GQ , Bai XJ. Analysis of physical composition and heavy metals pollution of municipal solid waste (MSW) in Beijing. IOP Conference Series: Earth and Environmental Science. 2018;**128**:012061. DOI: 10.1088/1755-1315/128/1/012061

[47] Perego A, Basile A, Bonfante A, De Mascellis R, Terribile F, Brenna S, et al. Nitrate leaching under maize cropping systems in Po Valley (Italy). Agritulture, Ecosystems and Environment. 2012;**147**:57-65. DOI: 10.1016/j.agee.2011.06.014

[48] Buczko U, Kuchenbuch RO, Lennartz B. Assessment of predictive quality of simple indicator approaches for nitrate leaching from agricultural fields. Journal of Environmental Management. 2010;**91**:1305-1315. DOI: 10.1016/j.jenvman.2010.02.007

[49] Crecchio C, Curci M, Mininni R, Ricciuti P, Ruggiero P. Short-term effects of municipal solid waste compost amendments on soil carbon and nitrogen content, some enzyme activities and genetic diversity. Biology and Fertility of Soils. 2001;**34**:311-318. DOI: 10.1007/s003740100413

[50] Weber J, Kocowicz A, Bekier J, Jamroz E, Tyszka R, Debicka M, et al. The effect of a sandy soil amendment with municipal solid waste (MSW) compost on nitrogen uptake efficiency by plants. European Journal of Agronomy. 2014;**54**:54-60. DOI: 10.1016/j.eja.2013.11.014

[51] Civeira G. Influence of municipal solid waste compost on soil properties and plant reestablishment in peri-urban environments. Chilean Journal of Agricultural Research. 2010;**70**(3):446-453

[52] Mkhabela MS, Warman PR. The influence of municipal solid waste compost on yield, soil phosphorus availability and uptake by two vegetable crops grown in a Pugwash sandy loam soil in Nova Scotia. Agriculture, Ecosystems and Environment. 2005;**106**:57-67. DOI: 10.1016/j.agee.2004.07.014

[53] Iglesias-Jiménez E, Álvarez CE. Apparent availability of nitrogen in composted municipal refuse. Biology and Fertility of Soils. 1993;**16**:313-318. DOI: 10.1007/BF00369312

[54] Jordán MM, García-Sánchez E, Almendro-Candel MB, Navarro-Pedreño J, Gómez-Lucas I, Meléndez I. Geological and environmental implications in the reclamation of limestone quarries in Sierra de Callosa (Alicante, Spain). Environmental Earth Sciences. 2009;**59**:687-694. DOI: 10.1007/s12665-009-0065-x

[55] Baldi E, Marcolini G, Quartieri M, Sorrenti G, Toselli M. Effect of organic fertilization on nutrient concentration and accumulation in nectarine (*Prunus persica var. nucipersica*) trees: The effect of rate of application. Scientia Horticulturae. 2014;**179**:174-179. DOI: 10.1016/j.scienta.2014.09.029

[56] Calleja-Cervantes ME, Fernández-González AJ, Irigoyen I, Fernández-López M, Aparicio-Tejo PM, Menéndez S. Thirteen years of continued application of composted organic wastes in a vineyard modify soil quality characteristics. Soil Biology and Biochemistry. 2015;**90**:241-254. DOI: 10.1016/j.soilbio.2015.07.002

[57] Liu A, Ren F, Lin WY, Wang JY. A review of municipal solid waste environmental standards with a focus on incinerator residues. International Journal of Sustainable Built Environment. 2015;**4**:165-188. DOI: 10.1016/j.ijsbe.2015.11.002

[58] Cheng Z, Sun Z, Zhu S, Lou Z, Zhu N, Feng L. The identification and health risk assessment of odor emissions from waste landfilling and composting. Science of Total Environment. 2019;**649**:1038-1044. DOI: 10.1016/j.scitotenv.2018.08.230

[59] Khalil C, Al Hageh C, Korfali S, Khnayzer RS. Municipal leachates health risks: Chemical and cytotoxicity assessment from regulated and unregulated municipal dumpsites in Lebanon. Chemosphere. 2018;**208**:1-13. DOI: 10.1016/j.chemosphere.2018.05.151

[60] Kjeldsen P, Barlaz MA, Rooker AP, Baun A, Ledin A, Christensen TH. Present and long-term composition of MSW landfill leachate: A review. Critical Reviews in Environmental Science and Technology. 2002;**32**(4): 297-336. DOI: 1064-3389/02/$.50

[61] Lim LY, Lee CT, Bong CPC, Lim JS, Sarmidi MR, Klemes JJ. A review on the impacts of compost on soil nitrogen dynamics. Chemical Engineering Transactions. 2018;**63**:349-354. DOI: 10.3303/CET1863059

[62] Rezapour S, Samadi A, Kalavrouziotis IK, Ghaemian N. Impact of the uncrontrolled leakage of leachate from a municipal solid waste landfill on soil in a cultivated-calcareous environment. Waste Management.

2018;**82**:51-61. DOI: 10.1016/j.wasman.2018.10.013

[63] Yusof N, Hraguchi A, Hassan MA, Othman MR, Wakisaka M, Shirai Y. Measuring organic carbon, nutrients and heavy metals in rivers receiving leachate from controlled and uncontrolled municipal solid waste (MSW) landfills. Waste Management. 2009;**29**:2666-2680. DOI: 10.1016/j.wasman.2009.05.022

[64] Zhao S, Zhang X, Duo L. Accumulation and spatial distribution of Cd, Cr and Pb in mulberry from municipal solid waste compost following application of EDTA and $(NH_4)_2SO_4$. Environmental Science and Pollution Research. 2013;**20**:967-975. DOI: 10.1007/s11356-012-0992-z

[65] Paradelo R, Villada A, Barral MT. Chemical fractionation of trace elements in a metal-rich amphibolite soil amended with municipal solid waste composts. Waste Biomass Valorization. 2018;**9**:1935-1943. DOI: 10.1007/s12649-017-9940-y

[66] Tella M, Bravin MN, Thuriès L, Cazevieille P, Chevassus-Rosset C, Collin B, et al. Increased zinc and copper availability in organic waste amended soil potentially involving distinct release mechanisms. Environmental Pollution. 2012;**212**:299-306. DOI: 10.1016/j.envpol.2016.01.077

[67] Vaverková MD, Elbl J, Radziemska M, Adamcová D, Kintl A, Baláková L, et al. Environmental risk assessment and consequences of municipal solid waste disposal. Chemosphere. 2018;**208**:569-578. DOI: 10.1016/j.chemosphere.2018.06.026

[68] Hamidpour M, Afyuni M, Khadivi E, Zorpas A, Inglezakis V. Composted municipal waste effect on chosen properties of calcareous soil. International Agrophysics. 2011;**26**:365-374. DOI: 10.2478/v10247-012-0051-5

[69] Adamcová D, Radziemska M, Ridosková A, Barton S, Pelcová P, Elbl J, et al. Environmental assessment of the effects of a municipal landfill on the content and distribution of heavy metals in *Tanacetum vulgare* L. Chemosphere. 2017;**185**:1011-1018. DOI: 10.1016/j.chemosphere.2017.07.060

[70] Page K, Harbottle MJ, Cleall PJ, Hutchings TR. Heavy metal leaching and environmental risk from the use of compost-like output as an energy crop growth substrate. Science of Total Environment. 2014;**487**:260-271. DOI: 10.1016/j.scitotenv.2014.04.021

[71] Gajski G, Orescanin V, Garaj-Vrhovac V. Chemical composition and genotoxicity assessment of sanitary landfill leachate from Rovinj, Croatia. Ecotoxicology and Environmental Safety. 2012;**78**:253-259. DOI: 10.1016/j.ecoenv.2011.11.032

[72] Lakhdar A, Rabhi M, GHnaya T, Montemurro F, Jedidi N, Abdelly C. Effectiveness of compost use in salt-affected soil. Journal of Hazardous Materials. 2009;**17**:29-37. DOI: 10.1016/j.jhazmat.2009.05.132

[73] Farrell M, Griffith GW, Hobbs PJ, Perkins WT, Jones DL. Microbial diversity and activity are increased by compost amendment of metal contaminated soil. FEMS. 2009;**71**:94-105. DOI: 10.1111/j.1574-6941.2009.00793.x

Chapter 4

Management of Organic Solid Waste in Meal Production

Luciléia Granhen Tavares Colares,
Gizene Luciana Pereira de Sales,
Aline Gomes de Mello de Oliveira
and Verônica Oliveira Figueiredo

Abstract

In Brazil, 31% of household food expenses are spent on meals eaten outside of the home. The food service sector is a major consumer of resources (water, energy, food, and other materials), and generator of solid waste (SW) food, being a focus of concern of national and international organizations, given their potential economic, social, and environmental impacts caused by the final disposal of solid waste. This work problematizes the generation of solid waste during the production of meals for collectivities and presents the study carried out in three community restaurants located in Rio de Janeiro, Brazil. The solid waste generated in all stages of the meal production process was weighed. Food leftovers and food scraps were the solid waste generated in greater quantity in the three restaurants studied, showing the need for better planning of the menus and the quantities of preparations produced, since these residues are closely related to the acceptance of the menu and to the waste of food. An organic solid waste management plan has been proposed based on environmental performance evaluation during large-scale meal production.

Keywords: organic solid waste, meal production, waste management, waste of food, community restaurants

1. Introduction

The increasing generation of municipal solid waste (MSW) is influenced by several factors such as population growth, urbanization, and lifestyle changes [1, 2].

The food service sector has contributed to the increase in solid waste (SW) generation, which corresponds to roughly 20% of all waste generated, as compared to households that account for approximately 50%. The final destination of this SW is therefore of great concern to municipalities and states [3].

According to the Brazilian Institute of Geography and Statistics, in Brazil, 31% of household food expenses are spent on meals eaten outside of the home [4], which boosts this economic sector as it generates direct and indirect jobs (210,000) and involves large sums of money (51 billion reals/year) related to the commercialization of foods and meals, as well as the consumption of significant quantities of fresh and processed foods [5]. However, the food sector contributes substantially to the production of solid waste, especially those that are organic-based.

The generation of organic solid waste (OSW) in the food service sector is closely related to food wastage. Although food losses and wastage are present throughout the food chain (production, harvesting, transportation, marketing, and consumption), large-scale meal production is a major factor contributing to food waste in the process flow due to menu planning that does not prioritize food from the crop, a lack of control of receipt and storage of food, and improper prepreparation practices and preparation, as well as other sources.

It is known that the solutions to the problems associated with solid waste generation in any human activity involve not only reduction at the source, but also the reuse or recycling of the discarded materials and the final environmental disposal of the tailings. For this, the efficient management of SW is necessary to contribute to more sustainable livelihoods [6].

The objective of this chapter is to problematize the generation of solid waste during the production of meals for collectivities and present a management plan that will contribute to the minimization of food waste and the environmental impacts caused by the final disposal of OSW.

2. Organic solid waste generated during the production of meals

In recent years, the Research and Extension Group on Sustainability in Meal Production (LASUPRE, Brazilian acronym) of the Federal University of Rio de Janeiro, Brazil, has studied the issue of food waste, the generation of OSW, and management strategies that can be employed during the production of large-scale meals in the food service sector [7–11]. Organic solid waste has been gaining prominence in research since it participates with more than 50% of the waste generated in all human activities in developing countries, and it is necessary to apply technology that reduces the losses and waste of food produced and consumed, as well as the final disposal of solid waste generated, which is environmentally sound.

Other important issues related to food loss and waste include resource use, solid waste generation, and energy-related emissions for transportation, storage, processing, and consumption, as well as wasted food calories, causing significant economic, social, and environmental impacts [3].

Albisu [12] emphasizes the need to differentiate the terms "food loss" and "food waste," since the former refers to the quantity of edible food that is not consumed after its harvest and the second is related to consumption itself. The seriousness of this issue is notorious, since 1.3 billion tons of food are wasted every year, while 800 million people are still starving [13].

According to Canali et al. [14], approximately a quarter of the total food calories produced globally are wasted, causing environmental impact related to the emission of greenhouse gases including approximately 3.3 GT of carbon dioxide (CO_2) equivalent.

The strategies for mitigating the problems of food loss and food waste are very important, yet different. While the first involves the supply chain from the production, harvesting, transportation, and distribution of food, the second involves the acquisition for individual or collective consumption. The Food and Agriculture Organization (FAO) [15] proposes various measures including (a) increasing the awareness of producers and consumers through information and communication campaigns, especially on agricultural and veterinary practices in the primary stages of production, as well as good manufacturing and hygiene practices in the preparation of food in the acquisition and consumption phases; (b) investing in small-scale agriculture, with training in processes and strategies for the conservation of harvested products, in addition to strengthening family farming through policies and programs on quality and safety; (c) improving transportation, energy, and marketing infrastructure, as well as developing technologies that contribute to the reduction of food losses and waste; (d) including education at all levels on the theme "Food and Nutrition Security (FNS)" and ways to avoid food losses and waste; and (e) encouraging South-South cooperation in all of the above measures.

Several countries have attempted to determine appropriate solutions for the final destination of organic solid waste such as incineration, anaerobic treatment with biogas, or energy generation and aerobic treatment with organic fertilizer production, in order to minimize the social, economic, and environmental impacts caused by the high generation of solid waste [16].

The Brazilian Solid Waste Policy (BSWP) emphasizes the difference between solid waste and tailings and recommends a hierarchy of solutions for solid waste management, starting with reduction at the generating source, treatment or recycling, and only then, the final disposal of tailings in an environmentally appropriate manner [17].

For the various stages of producing large-scale meals in the food service sector, sanitation controls have been established by legislation to prevent potential damage to food and distributed meals, thereby minimizing the risk of foodborne diseases [18–20]. However, certain documents already highlight the importance of assessing the environmental performance of organizations, taking into account the impacts of their activities [21] such as the high levels of water consumption and the generation of gases that contribute to the destruction of the ozone [22]. The inclusion of this theme in the production of meals is necessary since throughout the meal production process, resources (water, energy, food, and other materials), solid food, and nonfood solid waste are generated [10].

Figure 1 shows the production process flow of meals with their controls (process and hygienic sanitary) and the main types of solid waste that are generated. Colares and Figueiredo [8] evaluated the solid waste originating from a food service that produced and distributed 1500 meals a day. They found that 88% of the solid residues generated were organic waste, produced mainly in the prepreparation stage (34%), and during distribution of the meals (66%), which was represented by leftovers (meals produced and not distributed) and food scraps (distributed and nonconsumed meals). Similar to these results, when quantifying the solid waste generated in a university restaurant in the city of Maringá, Brazil, Zotesso et al. [23] found that over 21 d, 40,650 meals were served and 6.5 t of solid waste was produced (161 g of waste for each meal served). They observed high quantities of solid waste during the prepreparation stage (43%) from peeling and cutting, and food waste (47%). These results demonstrate the need for food service management to be more focused on minimizing food waste and the environmental impacts caused by poor solid waste management and disposal.

This work presents a study carried out on three community restaurants located in the city of Rio de Janeiro, Brazil, where a solid waste generation diagnosis was performed. A management plan that prioritizes the environmental performance evaluation during the production of meals in order to minimize food waste and associated environmental impacts is also proposed.

The Community Restaurants Program is a Food and Nutrition Security policy instrument that is implemented through a formal agreement between the Ministry of Social Development and Fight Against Hunger (MDS, Brazilian acronym) and the respective State. They promote the Human Right to Adequate Food (HRAF), especially for workers who purchase meals in the urban centers of the country. The municipalities are responsible for the administration and maintenance of the equipment [24].

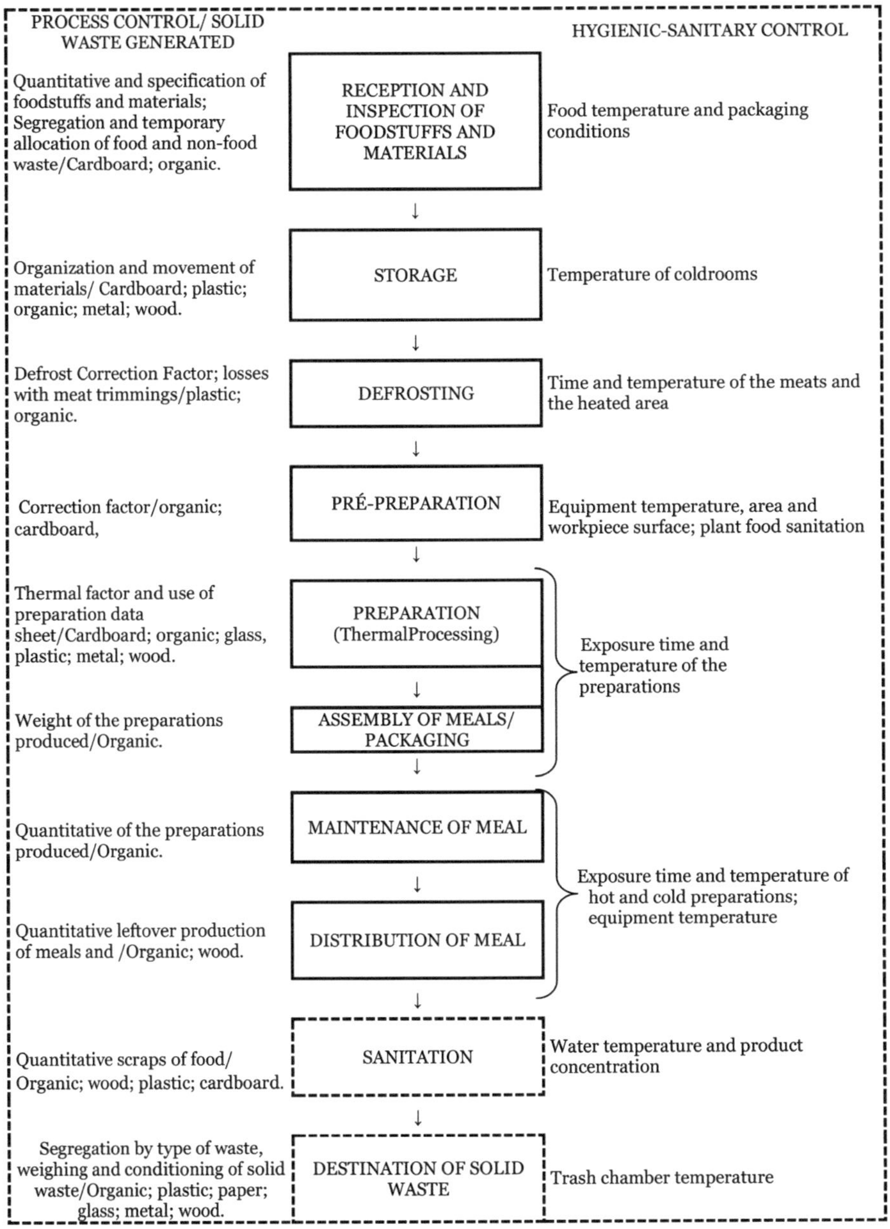

Figure 1.
Flowchart of the food production process with hygienic sanitary process controls and solid waste generation.

Currently, in the city of Rio de Janeiro, Brazil, there are three community restaurants in operation producing and serving approximately 6000 meals a day, including breakfast and lunch.

To implement these, restaurants, municipalities and states must meet the eligibility criteria specified in the call notice, which can be translated into geographical aspects. They must also have FSN or social assistance-related programs or instruments in place that will act in an integrated manner. The target audience is primarily low-income formal and informal workers, the unemployed, students, the elderly, and populations at social risk in urban centers and their peripheries. These restaurants should be deployed in high-volume areas of low-income workers such as the central areas of cities that are preferably close to mass transit [24].

3. Materials and methods

This quantitative cross-sectional study was conducted in three community restaurants in Brazil, and the methodology adopted in this study involves some steps.

3.1 Generation of OSW during the production and distribution of meals in community restaurants

The study was conducted over the period of 1 month in 2008 in three community restaurants located in the city of Rio de Janeiro (R1, R2, and R3), which together produced an average of 7500 meals/d for the socially vulnerable population, with the following menu: soup, salad, rice, beans, main course (protein), main course option, garnish, dessert (fruit or sweet), refreshment, bread, coffee, and tea.

The research was approved by the Research Ethics Committee of the Institute of Public Health Study of the Federal University of Rio de Janeiro (IESC-UFRJ) Opinion No. 03/2007, in compliance with Resolution 196/96 of the National Health requirements for conducting research with human beings.

3.1.1 Quantification and gravimetric composition of solid wastes generated

For 1 month, in the three community restaurants, the solid waste generated in all stages of the meal production process was weighed on a mechanical scale, model ruler, with a capacity of up to 150 kg and precision digital scale capacity of up to 5 kg (TGK-2261 model). A direct weighing technique was used (**Figure 2**). The mass balance was used in the prepreparation stage of the vegetables, whose peeling was done in the equipment "tuber peeler" [7]. The gravimetric composition of the generated residues was performed using Eq. (1).

$$\text{Gravimetric composition} = \frac{\text{weight of solid waste by type}}{\text{total weight of solid waste generated}} \times 100 \quad (1)$$

For the analysis of organic solid waste, the total production of the preparations to be served at meals, the leftovers (prepared and undistributed preparations) and the food scraps (preparations distributed and not consumed by the diners) were weighed. From this analysis, the percentage of leftovers (Eq. (2)) and index of food scraps (Eq. (3)) were calculated.

Figure 2.
Direct weighing of solid waste generated during the production of meals.

$$\text{Leftovers\%} = \frac{\text{Leftover weight}}{\text{Weight of meal produced}} \times 100 \quad (2)$$

$$\text{Index of food scraps} = \frac{\text{Weight of food scraps}}{\text{Distributed meal weight}} \times 100 \quad (3)$$

4. Characterization of organic solid waste in meal production

The amount of waste generated during the study period varied according to the number of meals produced in the three community restaurants, as can be seen in **Table 1**.

Among the solid residues, organics were generated in greater quantities, followed by recyclable residues (**Table 1**). Similar results were obtained by Zotesso et al. [23] and Colares and Figueiredo [8], who found that 82% and 88% of the residues produced were of organic origin, respectively.

Table 2 shows the gravimetric composition of the solid waste generated in the three restaurants. It was observed that of recyclable waste, there was more plastic, followed by cardboard. The carton came from boxes of primary, secondary, and tertiary packaging of meats, ice creams, tea, processed dessert, sugar, eggs, cups, and disposable napkins, among other materials.

The amount and type of waste generated in community restaurants were related to the planned menu and specificity of meal production.

Failure to plan the menu, caused by repeating meals in a short time and replicating cooking techniques on the same menu, can lead to nonacceptance by consumers, and consequently, food waste by the increase in leftovers and remains. **Table 3** presents the average amount of meals produced in the three community restaurants and the generation of residues from leftovers and food scraps.

At the menu planning stage, it is essential to use preventive measures that reduce the generation of solid waste. For example, using food directly from the harvest and diversifying preparation techniques can positively influence the formation of healthier eating habits by consumers, as well as a more sustainable menu through the reduction in food waste and solid waste [25, 26].

There was a failure at the menu planning stage in the community restaurants evaluated in this study. This occurred due to the presence of the same type of vegetable (potato) in two meals of the same menu (in the entree and in the garnish), which led to an excess in food remains that day (R1). Similarly, an excess of

Types of solid waste	Community restaurants					
	R1		R2		R3	
	kg	%	kg	%	kg	%
Recyclable	1772.0	9.2	1843.3	10.4	1254.8	10.3
Organic food waste	17,391.8	89.9	15,578.0	88.3	10,818.7	88.9
Residual waste	174.6	0.9	224.2	1.3	98.8	0.8
Frying oil	325.0	1.7	272.0	1.5	134.0	1.1
Total	19,338.5	100	17,645.5	100	12,172.3	100
Produced meals/month	62,700	—	60,000	—	51,000	—
RS per capita (Kg)	0.308	—	0.294	—	0.239	—

Table 1.
Solid waste generated and the number of meals produced per month in the three community restaurants.

Solid waste	Gravimetric composition					
	R1		R2		R3	
	kg	%	kg	%	kg	%
Paper	41.4	2.3	33.7	1.8	53.5	4.3
Cardboard	465.6	26.3	543.2	29.5	359.61	287
Wood	3.6	0.2	2.0	0.1	1.85	0.1
Plastic	1022.9	57.7	1058.3	57.4	741.37	59.1
Metal	191.7	10.8	160.5	8.7	96.1	7.7
Rigid plastic	46.8	2.6	45.8	2.5	2.4	0.2
Total recyclable	1772.04	100	1843.31	100	1254.83	100
Vegetable waste	2824.5	16.0	2106.4	13.3	2583.2	23.6
Meat waste	603.6	3.4	734.4	4.6	96.7	0.9
Coffee and tea drag	106.0	0.6	87.0	0.5	56.1	0.5
Food leftovers	5160.1	29.2	5843.5	36.9	3202.3	29.3
Food scraps	8697.6	49.2	6806.7	43.0	4880.3	44.6
Frying oil	292.5	1.7	244.8	1.5	120.6	1.1
Total organic food waste	17391.8	100	15,577.9	100	10818.7	100
Wet paper	102.2	58.5	146.1	65.2	44.0	44.5
Dirty paper	60.2	34.5	75.1	33.5	53.0	53.6
Sealing tape	12.2	7.0	3.04	1.4	1.8	1.8
Total residual waste	174.6	100	224.2	100	98.8	100
Total	19,338.5	—	17,645.5	—	12,172.3	—

Table 2.
Gravimetric composition of the solid waste generated in the three community restaurants over the period of 1 month.

Daily result	Average amount (kg) ± standard deviation (SD)					
	R1		R2		R3	
	kg	SD	kg	SD	kg	SD
Produced meals	3843.2	28.0	3054.7	331.3	2454.2	213.4
Food leftovers	271.6	118.8	292.2	178.3	188.4	72.5
Leftover %	7.1	3.0	10.1	7.6	7.6	2.8
Distributed meals	3571.6	328.8	2762.5	434.3	2272.6	199.1
Consumption per person	1.09	0.1	1.0	0.09	0.9	0.1
Food scraps	419.3	53.6	312.2	62.7	276.9	24.9
Index of food scraps	11.7	1.1	11.4	1.8	12.3	1.4

Table 3.
Weight of meals produced daily, meals consumed, leftovers, and food scraps.

sauce (mixed meat and meatballs of chicken) was observed in certain meals, resulting in a high quantity of leftovers and food scraps.

Regarding the frying oil, a high level of consumption was observed since the frequency of frying was quite high in the three community restaurants. Although contributing from 1.1 to 1.7% of organic solid waste (**Table 2**), during the data collection period, it was observed that this oil was used for many days before being changed, which constitutes an improper practice. It was also observed that on the day of the oil change, a large amount of metal residue was found in the fryer since the oil was conditioned in this vessel. In spite of this, the used oil was sent to a recycling company for biodiesel production.

A solution for reducing the production of these residues would be to diversify the preparation technique, for example, through the introduction of roasting techniques, which would reduce the need to fry the food [7].

In this present research, the factors that contribute to food waste were observed throughout the production process, and consequent generation of solid waste in the corresponding stages is shown in **Table 4**.

4.1 Receiving and storing raw materials

Failure to receive food and materials due to the lack of adequate physical space was observed in this study. Sometimes the various food types were received without proper inspection because they did not have adequate space.

After receiving, the raw materials or ingredients should be stored according to their perishability characteristics, under ambient temperature or in cold units such as cold rooms and freezers [18, 27].

The storage area was visibly inadequate in the three restaurants evaluated in this study, and there was excessive stacking of materials, which is in violation of the sanitary legislation [18]. These conditions were also verified by Ricarte et al. [27], who discussed food wastage as a result of inadequate storage conditions.

Solid waste	Stages of the meal production process					
	Reception and inspection of food stuffs and materials	Storage	Prepreparation of meats	Prepreparation of fruits and vegetables	Preparation/ cooking	Distribution
Paper	x	x				x
Cardboard		x	x		x	x
Wood						x
Plastic		x	x		x	x
Rigid plastic					x	
Metal					x	
Vegetable waste				x	x	
Meat waste			x			
Food leftovers					x	x
Food scraps						x
Frying oil					x	
Coffee and tea drag					x	
Wet paper					x	x
Sealing tape	x	x				

Table 4.
Identification of the solid waste generated at the various stages of the meal production process.

With respect to storage, cardboard-type residues were observed in this study, which came from the packaging of the food received. In order to minimize the generation of cardboard-type waste, food services should have an adequate area to receive inputs. This study showed that food was received on the sidewalk of the restaurant, which made it difficult to transfer the items from the carton to the plastic box.

Considering that cardboard was one of the most recyclable types of waste generated in the three food services (**Table 2**) and that this residue was present in several stages of the meal production process (**Table 4**), adequate processes or protocols for receiving raw food materials would aid in reducing it.

4.2 Prepreparation of meats and prepreparation of fruits and vegetables

The prepreparation stage is divided into two areas, one for the prepreparation of meat (poultry, fish, beef, and pork) and the other for the prepreparation of fruits and vegetables.

In the meat prepreparation stage, the following actions were carried out: clean, cut, or grind the meats. A low percentage of meat residues (skin, sebum, nerve, and bones) were observed in in the three community restaurants (**Table 2**).

The prepreparation of fruits and vegetables required the removal of damaged leaves, barks, and stalks, as well as cleaning and cutting. There was a higher level of residue generation from fruits and vegetables than that observed during the prepreparation of meats in the three restaurants involved in this study (**Table 2**). A similar result was obtained by Zotesso et al., who also observed greater volumes of organic solid residues from the fruits and vegetables in the prepreparation stage in a university restaurant [23].

Various factors may be related to food waste, and consequently, to solid waste generation in this stage, such as [7].

1. Poor quality of the raw material received;

2. Absence of preventive maintenance of the equipment (**Figure 3**);

3. Lack of training of food handlers to perform activities and reduce excessive removal of edible parts during the prepreparation of fruits and vegetables (**Figure 4**).

Andreatti et al. [28] and Ricarte et al. [27] also established that food was wasted during the prepreparation of fruits and vegetables in the restaurants that they

Figure 3.
Food waste due to the lack of equipment maintenance.

Figure 4.
Excessive removal of edible food parts.

studied. The authors indicated that the waste was related to the failure to receive materials and the procedure adopted for cutting, which involved excessive removal of barks and shavings.

These factors not only cause food wastage but also increase the cost of serving meals. Therefore, measures should be taken to adapt the menu, provide specific training for food handlers, monitor the activities carried out, and implement preventive and corrective maintenance of the equipment.

4.3 Preparation/cooking

At the cooking stage, heat was used in the preparation of food that was distributed to consumers. Food wastage was related to factors that led to low acceptance of meals by consumers, such as:

1. Preparation of large quantities of food at one time, causing a surplus in production;

2. Lack of monitoring of the activities performed, compromising the presentation of the meals;

3. Conditioning of meals that would be consumed hot, at room temperature;

4. Inadequacy of the use of the cooking technique, altering the sensorial characteristics of the meals.

It is worth noting that due to the inadequacy of the physical structure of the restaurants evaluated in this study, cardboard and plastic residues were generated (**Table 4**). This occurred because certain products such as industrialized breaded meatballs and chicken went from the storage area to the preparation area to be cooked with their secondary and tertiary packaging.

4.4 Distribution of meals

The organic waste generated at the distribution stage for the three community restaurants was related to the leftovers (meals produced and not distributed) (**Table 4**).

Vaz [29] acknowledged an acceptable level of leftovers of up to 3% since high percentages may indicate excess production, and consequently, food wastage. In this present study, the percentage of leftovers for all three restaurants was above the recommended level at 7.1, 10.1, and 7.6% for R1, R2, and R3, respectively (**Table 3**). Busato et al. [30] evaluated the percentage of leftovers in a community restaurant in Chapecó-Santa Catarina, Brazil, and observed that it was within the recommended level (1%). The authors emphasized the importance of leftover control in meal production, not only for assessing food waste and cost, but also as an indicator of the quality of the meal served and level of acceptance of the menu offered.

In contrast to Venzke [31], who identified the prepreparation area as one of the largest sources of organic waste, in the present research, it was observed that the largest generation of organic waste occurred at the distribution stage and was represented by leftovers and food scraps (**Table 2**).

Considering the quantity of leftovers from the restaurants under study (**Table 3**), 272, 292, and 188 additional daily meals could be served, respectively, at R1, R2, and R3. The excess of leftovers indicates a failure in the planning of the menu [32].

To minimize the quantity of leftovers, it is necessary to organize the work process, correctly plan for the number of meals and quantities per capita, monitor meal distribution with specific training for correct portioning, increase awareness, involve the team, and prepare smaller quantities where possible.

In this research, the following factors were observed related to the high levels of solid waste produced during the meal distribution stage:

1. Inadequate physical space, making it difficult to replace the meals on the distribution counter and compromising the presentation of the meals served to consumers;

2. Damaged equipment, making it difficult to maintain the temperature of the meals served;

3. Incorrect serving size of utensils, making it difficult to portion the meals;

4. Lack of care with exposure of the meals (**Figure 5**);

5. Inadequate planning with respect to the quantity of meals to be distributed.

Figure 5.
Lack of care with exposed meals.

Process flow step	PDCA cycle			
	Plan	Do	Check	Act
	Development of indicators	**Data collection and analysis**	**Critical review based on benchmarks**	**Improvement in environmental performance assessment**
Reception and inspection of food stuffs and materials	• Quantity of materials purchased per meal produced • Quantity of secondary and tertiary packaging per meal produced • Quantity of food discarded by week	• Buy only the necessary • Train workers • Reduce purchases with secondary and tertiary packaging • Recyclable Material Spreadsheet • Worksheet of non-conformities of genres • Comparison with patterns determined by the organization	• Control charts • Inspection during the process • Inspection of measuring Equipment • Audits and quality reviews Cost related to quality and solid waste management	• Establishment of targets for reduction of nonconformities: exchange of suppliers, visit to new suppliers, establishment of new standards • Integration between the waste management system and the management of hygienic sanitary quality and productive process of meals • Quantification and segregation of solid waste
Storage	• Quantity of food discarded by week	• Control the quantity of food stuffs necessary for food production • Shelf life of food • Suitable storage (under refrigeration and at room temperature)		
Prepreparation of meats	• Percent of defrosting meats • Quantity of meat shavings per kilo of meat • Number of portions produced by net weight			
Prepreparation of fruits and vegetables	• Amount of vegetable residues per volume processed • Edible parts indicator (correction factor)	• Control of the amount of vegetables necessary for the production of meals • Edible parts control worksheet (correction factor) • Number of operational stops due to equipment problems per week		

Process flow step	PDCA cycle			
	Plan	Do	Check	Act
	Development of indicators	**Data collection and analysis**	**Critical review based on benchmarks**	**Improvement in environmental performance assessment**
Preparation/cooking	• Water footprint per meal produced • Amount of water spent in the preparations • Number of preparations produced by number of meals planned • Number of operating stops due to problems with equipment for meals sold	• Control the water footprint of the planned menu • Equipment maintenance • Daily water consumption worksheet • Control of visible water losses • Training of workers for the rational use of water • Cooking temperature control of the preparations • Control of operational stops		
Distribution of meals	• Percentage of leftovers • Index of scraps of food • Quantity of leftovers per meal produced • Quantity of food scraps per meal sold • Queue time for the meal • Percentage of acceptance of preparations • Bounce rate per preparation	• Equipment maintenance • Training of workers for the rational use of water • Temperature control of preparations and equipment for maintenance and distribution • Stops control during meal distribution due to lack of preparation • Indicator control worksheet		

Table 5.
Management plan for organic solid waste generated in the production of meals.

Castro and Queiroz [33] indicated an ideal index of food scraps of less than 5%. In this present study, the incidence of remains above 10% detected at the three community restaurants (**Table 3**) may be related to planning failures such as monotony and repetitiveness in the supply of meals. Chamberlem et al. [34] identified potential causes of the generation of food scraps as the absence of new meals, monotony in the consistency of the meals served, inadequacy of the temperature of the prepared meals, and the above-standard portioning. In addition, inadequate ambient temperature and insufficient seating in the cafeteria may impact the amount of food wasted by consumers.

In the community restaurants assessed in this study, the separated OSW was stored in a refrigerated room before final disposal; however, the handling of the recyclable waste was carried out in an inadequate manner, as many residues were temporarily stored at room temperature in an uncovered area, and the cartons were mixed with plastic bags containing traces of blood from meat packaging. This form of waste handling and separation may lead to the proliferation of rodents and insects, in addition to making it difficult to recycle [7]. The better the separation, the greater the possibility for suitable treatment and/or disposal for the different types of waste generated [35].

In light of the above, the management of solid waste generated during the production of meals, especially the organics, is necessary to assist in the production of more sustainable meals.

5. Management processes

There have been many questions surrounding the issue of food waste throughout the food chain (from production to consumption), and some solutions are related to the improvement of food and solid waste legislation [36].

Organic solid waste management based on environmental performance evaluation during large-scale meal production can be an effective strategy to minimize the environmental, social, and economic issues arising from its generation [37]. As a proposal, based on the flow of the production process, it is possible to choose operational performance indicators using the Plan, Do, Check, Act (PDCA) cycle [38, 39] for the management of OSW, as shown in **Table 5**.

The prevention of food waste is the first priority in the proposed hierarchy of management of OSW, as emphasized by Andriukaitis [40]. In this sense, it is necessary to evaluate which losses are avoidable in order for the plan to be feasible. This is because inevitable losses must follow the second priority in the management hierarchy, which is the treatment (organic solid waste) or recycling (recyclable) of waste and will depend on the technology, infrastructure, incentives, finances, and markets that are available [40].

Finally, the management of OSW during the production of meals should include the planning of menus, logistics of supply of foodstuffs and materials, and the entire flow of production. These actions are critical since wasting water, energy, and calories accompany food waste.

6. Conclusions

The volume of food wasted during food production is of great concern and has implications for both food safety and the environment. Although food losses occur along the food production chain, it is necessary to study food waste both at the household and collective levels in restaurants (a specific segment of the food supply chain), in order to minimize this problem.

Community restaurants are a food safety mechanism utilized in the city of Rio de Janeiro, Brazil, designed to feed socially vulnerable populations. However, they still require management improvements in order to provide safe and sustainable food.

The assessment of solid waste generation in the three community restaurants allowed for the elaboration of a management plan based on the selection of environmental performance indicators for the various stages of the food production process, which facilitated the collection and analysis of comparable data with established goals.

It is therefore concluded that the management of restaurants cannot be isolated from the social, economic, and environmental issues resulting from the production of meals.

Acknowledgements

We would like to thank CAPES, which was especially helpful at all stages of the research providing the researchers with scholarships.

Conflict of interest

We declare that we have no conflict of interest of a financial, commercial, political, academic, or personal nature.

Author details

Luciléia Granhen Tavares Colares[1*], Gizene Luciana Pereira de Sales[2], Aline Gomes de Mello de Oliveira[1] and Verônica Oliveira Figueiredo[1]

1 Federal University of Rio de Janeiro, Rio de Janeiro, Brazil

2 Oswaldo Cruz Foundation, Bio-Manguinhos, Rio de Janeiro, Brazil

*Address all correspondence to: lucolares@nutricao.ufrj.br

References

[1] Rajpal A, Arora S, Bathia T, Bhargava R, Chopra AK, Kazmi AA. Co-treatment of organic fraction of municipal solid waste (OFMSW) and sewage by vermireactor. Ecological Engineering. 2014;**73**:154-161. DOI: 10.1016/j.ecoleng.2014.09.012

[2] Ali U, Sajid N, Khalid A, Riaz BL, Rabbani MM, Syed JH, et al. A review on vermicomposting of organic wastes. Environmental Progress & Sustainable Energy. 2015;**34**(4):1050-1062. DOI: 10.1002/ep.12100

[3] Scherhaufer S, Moates G, Hartikainen H, Waldron K, Obersteiner G. Environmental impacts of food waste in Europe. Waste Management. 2017;**77**: 98-113. DOI: 10.1016/jwasman.2018.04.038

[4] Instituto Brasileiro de Geografia e Estatistica—IBGE. Pesquisa de Orçamentos Familiares 2008-2009: Despesas, Rendimentos e Condições de Vida [Internet]. 2010. Available from: http://www.ibge.gov.br/home/estatistica/populacao/condicaodevida/pof/2008_2009/POFpublicacao.pdf [Accessed: August 1, 2016]

[5] Associação Brasileira das Empresas de Refeições Coletivas—ABERC. Mercado Real [Internet]. 2018. Available from: http://www.aberc.com.br/mercadoreal.asp?IDMenu=21 [Accessed: August 1, 2016]

[6] Chang NB, Davila E. Municipal solid waste characterizations and management strategies for the lower Rio Grande Valley, Texas. Waste Management. 2008;**28**:776-794. DOI: 10.1016/j.wasman.2007.04.002

[7] Sales GLP. Diagnóstico da geração de resíduos sólidos em restaurantes públicos populares do Município do Rio de Janeiro: contribuição para minimização de desperdícios. 2009. Dissertação (Mestrado em Nutrição) - Universidade Federal do Rio de Janeiro, Rio de Janeiro, 2009

[8] Colares LGT, Figueiredo VO. Gestão de resíduos sólidos gerados na produção de refeições. Revista Nutrição em Pauta. 2012;**114**:19-24. ISSN 1676-2274

[9] Mello A, Sales G, Jaeger L, Colares L. Estrutura físico-funcional de restaurantes populares do estado do Rio de Janeiro: influência sobre as condições higiênico-sanitárias. Demetra. 2013; **8**(2):91-101

[10] Zotesso J, Cossich E, Colares L, Tavares C. Analysis of solid waste generation in a university cafeteria in Brazil: A case study. Environmental Engineering and Management Journal. 2016;**15**(10):2327-2336. DOI: 10.30638/eemj.2016.254

[11] Colares L, Figueiredo V, Ferreira A, Oliveira A. Lista de verificação de boas práticas ambientais para serviços de alimentação: elaboração, validação de conteúdo e confiabilidade interavaliadores. Brazilian Journal of Food Technology. 2018;**21**:e2017066. DOI: 10.1590/1981-6723.06617

[12] Albisu L. Food losses and waste in the Spanish agro-food chain. CIHEAM Watch Letter. 2014;**30**:39-42. Available from: https://www.researchgate.net/publication/269277329_Tackling_Food_Losses_and_Waste_in_the_Mediterranean_from_knowledge_to_action [Accessed: November 18, 2018]

[13] Food and Agriculture Organization of the United Nations—FAO. The State of Food and Agriculture [Internet]. 2014. Available from: www.fao.org/publications [Accessed: November 18, 2018]

[14] Canali M, Amani P, Aramyan L, Gheoldus M, Moates G, Östergren K,

et al. Food waste drivers in Europe, from identification to possible interventions. Food Waste Drivers in Europe. 2017;**9**(37):1-33. DOI: 10.3390/su9010037

[15] Food and Agriculture Organization of the United Nations—FAO. Food Loss and Waste Facts [Internet]. 2015. Available from: http://www.fao.org/save-food/resources/infographic/en/

[16] Khalid A, Arshad M, Anjum M, Mahmood T, Dawson L. The anaerobic digestion of solid organic waste. Waste Management. 2011;**31**(8):1737-1744. DOI: 10.1016/j.wasman.2011.03.021

[17] Brazil. Lei No 12.305, de 2 de agosto de 2010. Institui a Política Nacional de Resíduos Sólidos que altera a Lei no 9.605, de 12 de fevereiro de 1998 e dá outras providências. Brazil: Diário Oficial da União. 2 de agosto de 2010

[18] Brazil. Agência Nacional de Vigilância Sanitária. Resolução da Diretoria Colegiada—RDC No 216, de 15 de setembro de 2004. Estabelece procedimentos de Boas Práticas para serviços de alimentação a fim de garantir as condições higiênico-sanitárias do alimento preparado. Brazil: Diário Oficial da União 2004; 17 set.; 2004

[19] Associação Brasileira de Normas Técnicas—ABNT. ABNT NBR 15635: 2015—Food services—Requirements of Good Practices Hygienic-Sanitary and Essential Operational Controls; 2015

[20] World Health Organization—WHO, Food and Agriculture Organization of the United Nations—FAO. Codex Alimentarius: Hygiene Basic Texts; 2009

[21] Associação Brasileira de Normas técnicas—ABNT. ABNT NBR ISO 14031. Gestão ambiental—Avaliação de desempenho ambiental: Diretrizes; 2013. 38

[22] Waste Resource and Action Programme—WRAP. New estimates for house hold food and drink waste in the UK [Internet]. 2011. Available from: http://www.wrap.org.uk/content/new-estimates-household-food-and-drink-waste-uk [Accessed: November 10, 2018]

[23] Zotesso JP, Cossich ES, Colares LGT, Tavares CG. Avaliação do desperdício de alimentos e sua relação com a geração de resíduos em um restaurante universitário. ENGEVISTA. 2016;**16**(2):294-308. DOI: 10.22409/engevista.v18i2.730

[24] Ministério do Desenvolvimento Social e Combate à Fome. Restaurantes Populares [Internet]. 2012. Available from: https://www12.senado.leg.br/cidadania/edicoes/388/conheca-o-programa-restaurantes-populares-do-ministerio-do-desenvolvimento-social-e-combate-a- fome [Accessed: October 7, 2018]

[25] Pistorello J, Conto SM, Zaro M. Geração de resíduos sólidos em um restaurante de um Hotel da Serra Gaúcha. Engenharia Sanitária Ambiental. 2015;**20**(3):337-346. DOI: 10.1590/S1413-41522015020000133231

[26] Strasburg VJ, Jahno VD. Paradigmas das práticas de gestão ambiental no segmento de produção de refeições no Brasil. Engenharia Sanitária Ambiental. 2017;**22**(1):3-12. DOI: 10.1590/S1413-41522017155538

[27] Ricarte M, Fé M, Santos I, Lopes A. Avaliação do desperdício de alimentos em uma unidade de alimentação e nutrição institucional em Fortaleza-CE. RESC. 2008;**1**(1):158-175

[28] Andreatti JAN, Bernardi M, Abbud RMR. Estudo do desperdício de vegetais no processo de pré-preparo de refeições em uma unidade de alimentação e nutrição. Revista FUNEC Científica—Faculdade Integrada de Nutrição

Santa Fé do Sul. 2013;1(1):1-10. ISSN 2318-5287

[29] Vaz CS. Restaurantes—Controlando Custos e Aumentando Lucros. 2006. 193 p. ISBN: 85-7238-228-3

[30] Busato MA, Barbosa FM, Frares KR. A geração de sobras e restos no restaurante popular de Chapecó (SC) sob a ótica da produção mais limpa. Revista Simbio-Logias. 2012;**5**(7):23-33 http://www.ibb.unesp.br/Home/Departamentos/Educacao/Simbio-Logias/a_geracao_sobras_restos_no_restaurante.pdf

[31] Venzke C. A geração de resíduos em restaurantes analisada sob a ótica da produção mais limpa. XXI Encontro Nacional de Engenharia de Produção, 2001, Salvador BA Anais do XXI Encontro Nacional de Engenharia de Produção Salvador 2001. ISSN ENEGEP: 2594-9713

[32] Teixeira SMFG, Oliveira ZMCO, Rego JC, Biscontini TM. In: Atheneu, editor. Administração aplicada às unidades de alimentação e nutrição. Rio de Janeiro; 2007. ISBN: 85-7379-251-5

[33] Castro F, Queiroz V. Cardápios: planejamento e etiqueta. 2007. 91 p. ISBN: 978-85-7269-333-2

[34] Chamberlem SR, Kinasz TR, Campos MPFF. Resto de ingestão e sobra descartada—fonte de geração de resíduos orgânicos em unidades de alimentação e nutrição em Cuiabá—MT. Alimentação e Nutrição Araraquara. 2012;**23**(3):317-325 http://fcfar.unesp.br/ojs-2.1.0-1/index/index.php/alimentos

[35] Falqueto E, Kligerman DC. Gerenciamento de resíduos oriundos da fabricação e distribuição do medicamento Diazepan para o município de São Mateus, ES. Ciência e Saúde Coletiva. 2008;**13**(supleme): 673-681. DOI: 10.1590/S1413-81232008000700016

[36] Stenmarck A, Jensen C, Quested T, Moates G. Estimates of European Food Waste Levels. Report of the Project FUSIONS [Internet]. 2016. ISBN: 978-91-88319-01-2. Available from: https://www.eu-fusions.org/phocadownload/Publications/Estimates%20of%20European%20food%20waste%20levels.pdf [Accessed: February 17, 2019]

[37] Associação Brasileira de Normas técnicas—ABNT. ABNT NBR ISO 14015: Gestão ambiental—Avaliação ambiental de locais e organizações (AALO); 2003. 18

[38] Sangpikul A. Implementing academic service learning and the PDCA cycle in a marketing course: Contributions to three beneficiaries. Journal of Hospitality, Leisure, Sport & Tourism Education. 2017;**21**:83-87. DOI: 10.1016/j.jhlste.2017.08.007

[39] Sokovic M, Pavletic D, Pipan KK. Quality improvement methodologies—PDCA cycle, RADAR matrix, DMAIC and DFSS. Journal of Achievements in Materials and Manufacturing Engineering. 2010;**43**(1):476-483

[40] Andriukaitis V. Keynote Speech at REFRESH Food Waste 2017. EU Comm, Comm Health Food Safety Berlin [Internet]. 2017. Available from: https://eu-refresh.org/sites/default/files/REFRESH%202017%20Conference_Minutes%20and%20Proceedings_jun2017.pdf#overlay-context=conference2017 [Accessed: October 7, 2018]

Chapter 5

Municipal Solid Waste Management and the Inland Water Bodies: Nigerian Perspectives

Akindayo A. Sowunmi

Abstract

Municipal solid waste (MSW) composition, natural transformation, dynamics and impacts on inland water bodies in Nigeria were examined, using dumpsites and landfills as the common markers. Nigeria is estimated to have over 178.5 million people and kg/capita/day of 0.26–1.02 MSW, projected to increase with the expansion of the economy which is in need of better articulated MSW management strategies. The enormous natural inland surface and groundwater resources are daily challenged directly and indirectly, through decline in physical, chemical and biological quality. Solid waste disposal along the waterways and leachates from natural activities on materials at dumpsites and landfills was strongly identified and recognized as the source of pollutant inputs. The immediate and projected public health consequences in changes in inland waters were provided for resident aquatic organisms, some of which serves as food for resident human populations that are largely dependent on these water bodies for their daily water requirements.

Keywords: Nigeria, inland water bodies, municipal solid wastes, water quality, public health

1. Introduction

Municipal solid wastes (MSW) refer to all wastes generated, collected, transported and disposed of within the jurisdiction of a municipal authority. In most cases, it comprises mainly food waste, discarded materials from residential areas, street sweepings, commercial and institutional nonhazardous wastes as well as (in some countries) construction and demolition waste.

MSW has been variously described as aggregation of unwanted materials generated from a range of human-related activities denominated from domestic to production. The origin of what is regarded as MSW can be closely associated with the earliest attempts by humans to transit from migrant to settler modes of living, which imposed the need to modify or change the character of raw or primary materials available to support or sustain the new modes of living and originating human activity.

Nigeria is the dominant country in West Africa, accounting for 47% of West Africa's population, with gross domestic product (GDP) growth at an average rate of 5.7% per year between 2006 and 2016, facilitated by volatile oil prices to a highest of 8% in 2006 and lowest of −1.5% in 2016; Human Development Index value also increased by 13.1% between 2005 and 2015 [1]. However, the country continues to face massive developmental challenges including, but not limited to, human development indicators and the living conditions of the population. Last collected in 2012 by the Nigeria National Bureau of Statistics, the total population of citizens in Nigeria was around 166.2 million people. In 2016, it was estimated to have over 178.5 million people although the United Nations' projections have placed the population as high as 186 million.

While MSW is generally associated with urbanization, recent developments in manufacturing processes have lowered the cost of production, enhancing the ability of manufacturers to produce goods that captures different income groups in population. The resultant effects are that areas hitherto considered as rural areas now experience both technological and economic penetrations. These penetrations will be accompanied by the penetration of MSW problems, hitherto restricted to urban centres. The developmental pressure experienced by major Nigeria cities has precipitated the upsurge in establishment of satellite towns, with attendant increase in human activity range and hence of waste generation.

Nigeria is considered one of the countries endowed with appreciable natural water resources in the world with the presence of the Niger River which is the third largest in Africa [2]. Natural water resources in Nigeria include enormous yearly rainfall, large surface bodies of water of rivers, streams and lakes, as well as in abundant reservoirs of underground water whose extent and distribution have not been fully assessed. The country is well drained with a reasonably close network of rivers and streams (**Figure 1**). Some of these rivers, particularly the smaller ones, are, however, seasonal, especially in the northern parts of the country where the rainy season is only 3 or 4 months in duration. In addition, there are natural water bodies like lakes, ponds as well as lagoons, particularly in the coastal areas [3–5]. The hydrology of Nigeria is dominated by two great river systems, the Niger-Benue and the Chad systems. With the exception of a few rivers that empty directly into the Atlantic Ocean (Cross River, Ogun, Oshun, Imo, Qua Iboe and a few others),

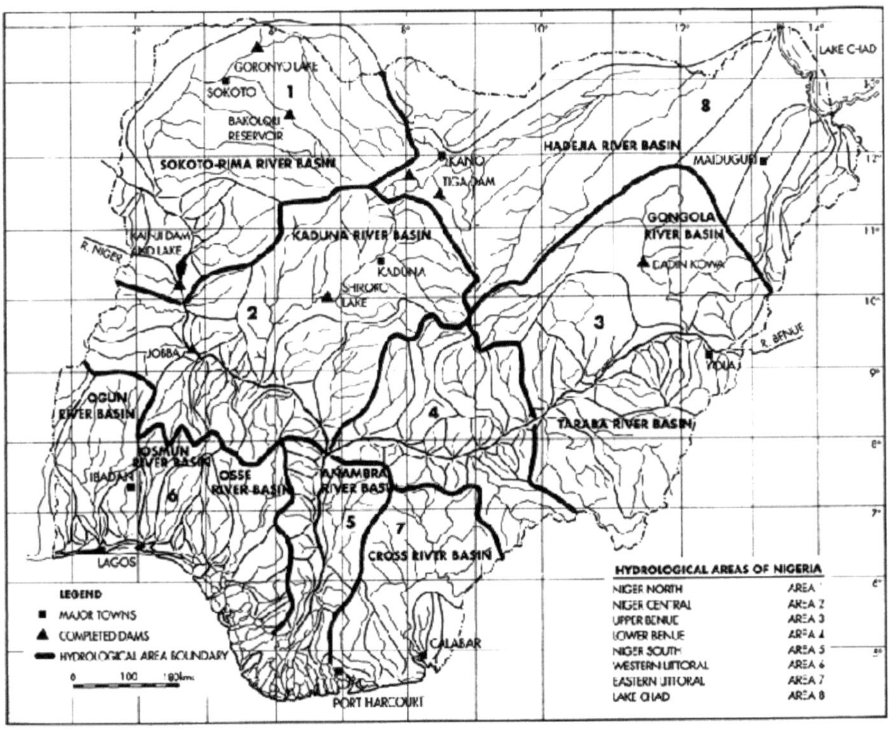

Figure 1.
Inland surface water resources of Nigeria (Source: [3, 4]).

all other flowing waters ultimately find their way into the Chad Basin or down the lower Niger to the sea. Nigeria lies between longitudes 2° 49′E and 14° 37′E and latitudes 4° 16′N and 13° 52′ North of the equator. The climate is tropical, characterized by high temperatures and humidity as well as marked wet and dry seasons, though there are variations between south and north. Total rainfall decreases from the coast northwards. The south (below latitude 8°N) has an annual rainfall ranging between 1500 and 4000 mm and the extreme north between 500 and 1000 mm.

The country has a vast expanse of inland freshwater and brackish ecosystems with an extensive mangrove ecosystem of which a great proportion lies within the Niger Delta. Freshwaters start at the northern limit of the mangrove ecosystems and extend to the Sahelian region. The major rivers, estimated at about 10,812,400 hectares, make up about 11.5% of the total surface area of Nigeria which is estimated to be approximately 94,185,000 hectares. Lakes and reservoirs have a total surface area of 853,600 ha and represent about 1% of the total area of Nigeria. Thus the total surface area of water bodies in Nigeria, excluding deltas, estuaries and miscellaneous wetlands, is estimated to be about 14,991,900 ha or 149,919 km^2 and constitutes about 15.9% of the total area of Nigeria. This review provided an insight on interactions between MSW, as indexed by dumpsites and landfills in Nigeria, and inland surface and groundwater in their vicinity.

2. Municipal solid waste generation in Nigeria

Generation of MSW in Nigeria is a daily occurrence, arising from diverse and varied human activities; hence the character of solid waste generated is never homogenous (**Plate 1a–k**). The differences can be a function of several indicators which include but not limited to originating tasks, income bracket, location, population density, population characteristics, culture, consumption pattern and seasons [6]. The quantity of MSW generated across cities in Nigeria is closely associated with population, economic, political and commercial activities. All these variables are however tied to the human element, as the driver of these changes. Changes in population pattern have been closely associated with changes in waste generation, even in the presence of optimally articulated management approaches. **Table 1** presented the close relationship between population and waste generation for the world's regions. It is noteworthy that regions with increasing or high per capita for MSW are the regions with high income. Nigeria had a population increase of between 2.6 and 2.7% annually between 2010 and 2018 and oscillated around 2% since 1965 [7]. The per capita income also increases steadily [1] which translated to increased purchasing power and consumptions of more products, with attendant waste generation. However, apart from Lagos State, waste generation data are not readily available or limited in coverage. The MSW per capita per day for different Nigerian cities is presented in **Table 2**, while **Figure 2** showed MSW per capita for low- (Agric), middle- (Bariga and Ojodu) and high-income (Lagos Island) locations in Lagos State over a period of 30 days. These values are comparable to the suggested per capita for the African region (**Table 1**).The data presented further extended the suggestion of a direct relationship between economic success and waste generation.

Accordingly, solid waste can be classified into four different types [13] depending on their source, which include:

a. Household waste, generally classified as municipal waste.

b. Industrial waste, as hazardous waste.

c. Biomedical waste or hospital waste, as infectious waste.

d. Electronic waste (e-waste).

Plate 1.
Selected sources and disposal of municipal solid wastes in Nigeria. (a) Waste paper, (b) Cassava peels and chaff, (c) mixed wastes, (d) waste labels, (e) waste bottles, (f) sorted wastes, (g) dumpsite in Sango-Ota (Ogun State), (h) dumpsite in Benin (Edo State), (i) Kara abattoir (Ogun State), (j) Ona River (Oyo State), (k) Mile 2 canal (Lagos State), and (l) Ikpoba River (Edo State).

Region	Current available data			Projections for 2025			
	Total urban population (millions)	Urban waste generation		Projected population		Projected urban waste	
		Per capita (kg/capita/day)	Total (tons/day)	Total population (millions)	Urban population (millions)	Per capita (kg/capita/day)	Total (tons/day)
AFR	260	0.65	169,119	1152	518	0.85	441,840
EAP	777	0.95	738,958	2124	1229	1.5	1,865,379
ECA	227	1.1	254,389	339	239	1.5	354,810
LCR	399	1.1	437,545	681	466	1.6	728,392
MENA	162	1.1	173,545	379	257	1.43	369,320
OECD	729	2.2	1,566,286	1031	842	2.1	1,742,417
AR	426	0.45	192,410	1938	734	0.77	567,545

Source: World Bank Group [1].
AFR, Africa Region; EAP, East Asia and Pacific Region; ECA, Europe and Central Asian Region; LCR, Latin America and Caribbean Region; MENA, Middle East and North Africa Region; OECD, Organisation for Economic Co-operation and Development; SAR, South Asia Region.

Table 1.
Current and projected generation pattern for different regions of the world.

City	Kg/capita/day
Lagos	0.63
Kano	0.56
Ibadan	0.51
Kaduna	0.58
Port Harcourt	0.60
Makurdi	0.48
Onitsha	0.53
Nsukka	0.44
Abuja	0.45–0.74
Ado Ekiti	0.71
Akure	0.54
Abeokuta	0.60–0.66
Aba	0.46
Ilorin	0.43
Lafia	0.39–1.02
Gombe	0.26–0.29
Makurdi	0.37–0.62

City	Kg/capita/day
Jimeta	0.39–1.02
Gboko	0.41–0.49

Source: Refs. [8–11].

Table 2.
Per capita wastes for Nigerian cities.

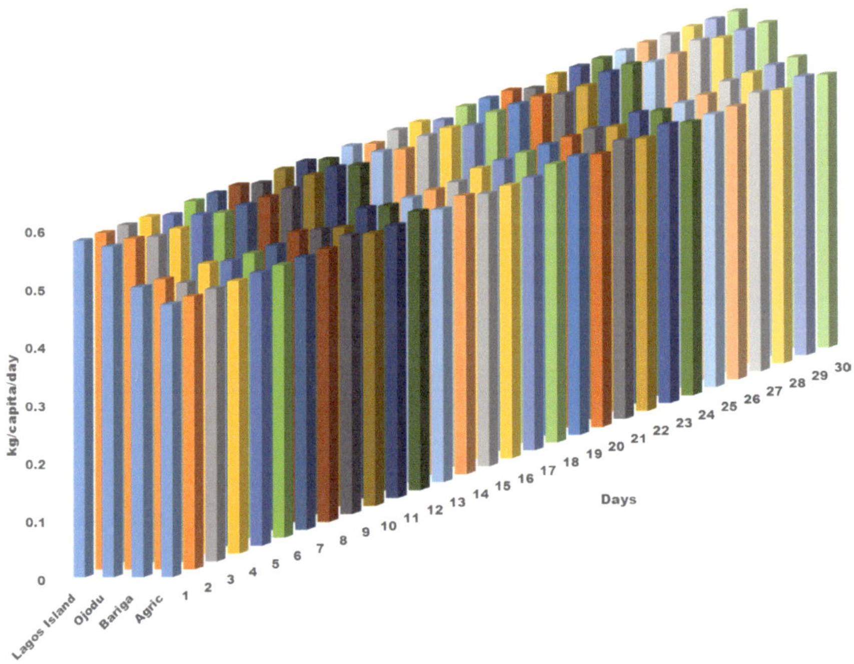

Figure 2.
Daily per capita waste generation of different incomes and densities from Lagos State, Nigeria. Modified from [12].

It is important to mention that until recently in Nigeria, MSW disposal methods (**Table 3**) received very little attention because wastes were considered an entity with homogenous properties [26, 27] or largely dominated by organic/decomposable wastes. Previous reports [28, 29] clearly supported this position and also suggested that study of wastes in Nigeria started in the 1970s. The components of MSW from different parts of Nigeria are presented in **Tables 4–8**, which showed that MSW are still largely dominated by organic/decomposable components. Shift in waste characteristics is however gradually becoming apparent reflecting changes from previously ignored traditional household electronic wastes to high-profile ubiquitous wastes of a technology-driven economy, in the form of heterogeneous components, popularly referred to as electronic wastes (e-wastes) and related components. The wastes from traditional household electronics have also increased with better purchasing power over time.

E-wastes were largely unacknowledged in Nigeria and considered part of MSW until the Koko waste incidence of 1988. This led to the separation of discarded household, ICT and personal electronic devices as e-wastes [48, 49]. The availability of cheaper versions of everyday ICT and personal electronic devices now provides additional source of consistent waste volume arising from short life cycle of substandard products. The volume and characteristics of MSW showing

Collection methods	Aba	Abeokuta	Abuja	Akure	Akoko Edo	Gombe	Birnin Kebbi	Maiduguri	Makurdi	Yola	Ughelli	Benin
Waste collection contractor	X	X	X	X		X		X	X		X	
Deposited at waste dump	X	X	X	X	X	X	X	X	X	X	X	X
Solid to other industries/ recycling	X	X	X					X	X		X	
Deposited in the river	X	X			X							
Deposited in drainage		X		X	X	X				X		
Compositing	X		X						X		X	
Incinerating/burning	X	X	X	X	X	X	X	X	X	X	X	X
Burying	X	X	X		X			X	X	X	X	X
Open space/plot dumping		X			X	X	X	X	X	X	X	X
Government trucks		X	X	X	X			X	X		X	

Modified from [8–10, 14–25].

Table 3.
MSW disposal methods from different Nigeria cities.

Waste components	Aba	Abeokuta	Maiduguri	Gombe	Ilorin	Kano	Warri
	Abia State	Ogun State	Borno State	Yobe State	Kwara State	Kano State	Delta State
Rubber						11.30–18.50	
Plastics	6.25	24.95	18.10		23.00–27.80		2.35–4.89
Papers	9.90	25.57	7.50	22.00–26.00		3.84–23.55	0.48–4.19
Glass	4.69	5.75	4.30	20.00–24.00	12.00–26.10	2.75–20.55	4.16–10.41
Aluminum scraps	9.90					2.20–9.49	0.52–4.69
Metal scraps	10.41	5.26	9.10				27.25–31.01
Tins and cans							8.71–20.71
Ceramics							0.35–3.74
Wood							1.19–4.39
Textiles		9.48	3.90			3.80–9.30	0.39–2.84
Compostable (e.g. food and wood)	47.39		25.80				
Food waste only				28.00–32.00	24.00–30.90	4.20–31.56	
Leaves and human feces					6.50–14.10		
Vegetables						13.30–23.00	0.26–7.62
Water sachets and cellophane packages	11.45						4.99–9.08
Hazardous wastes		2.69					
Ash			21.50			1.10–22.54	
Miscellaneous/others			9.80	22.00–28.00	15.60–21.00	1.74–6.35	20.71–34.91
	[15]	[8]	[23]	[24]	[30]	[31]	[32]

Table 4.
Waste components from Nigerian cities. I.

Waste components	Kaduna	Zaria	Onitsha	Yenagoa	Yola	Jos	Gboko	Makurdi
	Kaduna State		Anambra State	Bayelsa State	Adamawa State	Plateau State	Benue State	
Rubber	35	36	10.1	20.7–24.6				
Plastics			17.9		18.3	6.2–7.89		
Papers			8.1	13.6–14.7		17.7–22.3	10.0–14.0	2.1–10.9
Glass			4.5	9.4–10.9	3.0	7.9–13.1	7.0–10.0	0.1–6.9
Aluminum scraps								
Metal scraps			8.7	5.7–5.9	5.8	6.3–7.5	9.0–11.0	0.7–3.4
Tins and cans								
Water sachets and cellophane packages						6.3–9.9	15.0–22.0	5.9–10.2 (+plastics)

Waste components	Kaduna	Zaria	Onitsha	Yenagoa	Yola	Jos	Gboko	Makurdi
	Kaduna State		Anambra State	Bayelsa State	Adamawa State	Plateau State	Benue State	
Ceramics								
Textiles	5	1	10.1		67.6	5.7–8.6	9.0–12.0	0.3–6.1
Wood	16	26	10.1	2.5–3.6		5.5–12.6		
Compostable (e.g. food and wood)			6.5				15.0–21.0	23.4–57.5
Food waste only			40.50	40.8–42.8		12.2–14.2		
Vegetables						13.4–15.2		
Leaves and human feces	29	19						
Hazardous wastes	20	14						
Leather						3.8–6.6		
Ash/fines				2.1–2.8			10.0–12.0	21.0–48.7
Miscellaneous/others			3.7		5.3		8.0–10.0	1.7–28.9
	[33]		[34]	[35]	[36]	[37]	[38]	[21]

Table 5.
Waste components from Nigerian cities. II.

Waste components		Abuja	
Rubber	0.2–3.4		8.1–26.7
Plastics	16.2–21.3	3.4	2.3–13.9
Papers	6.9–13.6	25.3	3.2–13.4
Glass	4.1–5.5	3.00	0.8–6.5
Metal scraps	3.3–6.7	3.14	1.0–7.9
Tins and cans			
Ceramics			0.1–8.8
Textiles	0.1–4.7	3.0	0.2–4.8
Compostable (e.g. food and wood)		42.6	44.1–65.1
Food waste only	52.0–65.3		
Leaves and human feces			
Vegetables			
Water sachets and cellophane packages		14.5	7.8–18.6
Hazardous wastes		2.8	1.1–5.5
Ash			1.0–10.7
Miscellaneous/others	0.6–2.8	2.2	0.9–11.2
	[10, 25]	[16]	[39]

Table 6.
Waste components from Abuja.

Waste components	Lagos				
Plastics	7.29	3.6	5.0	15	6
Papers	10.2	12.5	10.0	10	6
Glass	2.8	1.8	2.0	5	8
Aluminum scraps					
Metal scraps	4.1	2.1	3.0	5	10
Water sachets and cellophane packages		7.7	9.0		
Textiles	3.8		5.0	4	6
Compostable (e.g. food and wood)	29.8	68.2		8	8
Food waste only			66.0		
Leaves and human feces					
Vegetables				45	50
Bones	1.8				
Ash/fines	21.2	4.2		8	10
Miscellaneous/others	18.8				
	[40]	[41]	[12]	[42]	[43]

Table 7.
Waste components from Lagos State.

Waste components	Port Harcourt				
Rubber					7.6
Plastics	1.5–8.3	2.2–4.8	11.5	18.0	9.9
Papers	4.0–16.5	5.6–16.5	12.3	24.2	12.4
Glass	0.2–6.3	0.2–2.5	9.5	10.9	13.5
Metal scraps	0.5–15.0	0.5–4.0	15.2		17.2
Tins and cans				10.9	
Water sachets and cellophane packages	9.9–18.5	10.5–14.7			
Textiles					7.6
Wood				18.0	8.4
Compostable (e.g. food and wood)	52.1–69.0	60.0–69.0	51.5		
Food waste only					29.2
Leaves and human feces					
Vegetables				18.0	
Miscellaneous/others	2.0–8.1	2.0–6.8			1.8
	[11]	[44]	[45]	[46]	[47]

Table 8.
Waste components from Port Harcourt.

e-waste proportion from dumpsites or landfills were absent from available studies. The isolation of e-waste as a unique recent component, activities of scavengers or pickers, electronic market dumpsites and dedicated studies to e-wastes probably contributed to the lack of such data.

3. Municipal solid waste and inland water bodies in Nigeria

The magnitude of changes experienced by inland water bodies as a result of MSW in Nigeria could be attributed to inappropriate siting, design, operation and maintenance of dumps and landfills. The history of the association between changes in quality of inland waters and MSW generation in Nigeria has not been adequately documented. However, classical reports [50, 51] provided a different trajectory to the narratives, where low-level perturbations reported for both the Ona River and Ogunpa River were associated with generation and disposal of MSW in Ibadan, Southwest Nigeria. Inland waters in Nigeria have been on the receiving end of MSW, but the details have been patchy. Inland surface and groundwaters in the vicinity of dumpsites in Nigeria have been reported to be generally compromised, and leachates have been the most cited reason.

Dumpsites usually undergo modification of wastes [52] in the following five basic steps:

- Phase I (lag phase/initial adjustment).
- Phase II (transition phase).
- Phase III (acid formation phase).
- Phase IV (methane production/fermentation phase/methanogenic phase).
- Phase V (maturation phase).

The products of these processes include volatilized chemicals as gas, leachate and changing community of organisms, all of which have profound influence on the physical, chemical and biological conditions in the immediate surroundings.

Leachates from dumpsites and landfills have been characterized (**Table 9**) and associated with contamination of inland surface water (**Table 10**) and groundwater (**Table 11**) resources from different parts of Nigeria with profound physical, chemical and biological consequences. Aquatic life and recreational criteria [53] suggested compromise in physical and chemical qualities, due largely to the presence of dumpsites close to these water bodies. Also age and the unique composition or characteristics of wastes deposited at dumpsites will greatly influence the resultant water quality. The biotic or biological responses of resident organisms to changes as elicited by activities associated with dumpsites have not attracted deserved attention or investigation considering the ecological and public health consequences. However, limited laboratory studies on aquatic organisms, *Chironomus* sp. *Culex pipiens*, *Bufo regularis* tadpoles and *Clarias gariepinus*, using products from dumpsites in the form of leachates from Oyo [57, 58, 77] and Lagos [77, 78] States showed pronounced aberrant behavioral responses and gross morphological and genetic damages. In spite of the limited studies from Nigeria, the reports agreed with comparable reports from other parts of the world on the negative influence of products of dumpsites on surface inland waters.

Groundwater in Nigeria provides water supply for 40.1% of Nigerians [79] and is considered to be the preferred source of water for different sectors providing about 40% of water public water supply [80] underlying the importance of groundwater sources. The integrity of such groundwater is therefore of importance because of direct consequences on human health. The quality of groundwater showed the presence of substances considered dangerous to human health at concentrations above standards [53, 68] considered acceptable. The detection of cadmium, nickel,

Parameters	National criteria	Odo Oba (Osun State)	Aba Eku (Oyo State)		Olusosun (Lagos State)		Aba Eku	Olusosun	Aba Eku
			Raw	Simulated	Raw	Simulated			
pH	6.5–8.5	6.25	8.6	4.9–5.5	7.30	6.80	7.8	8.1	8.0–8.3
Color		Dark brown					Dark brown	Dark brown	
TS		5072.17	3054.50	3281.00–4206.00			3116.67	4100.3	433–2091
TSS	0.75		1085.00	220.00–2490.00					1–460
TDS		3400	1969.50	1716.00–3412.00	0.32	1.32			
Total hardness		259.36			540		532	615	
Chloride		30	42.00	34–38	770.00	240.00	1106	1099	149–4280.0
BOD	6.0		3.67	2.99–3.83	598.00	590.00	601	594	110.7
COD	30.0		5.50	4.50–5.78	480.00	370.00	512	487	29–338.2
Turbidity			1030.00	440–1875					
Phosphate	3.5		895	175.5–450.73			122.02	215.7	ND
Nitrate	40.0		97.94	24.92–170.84	3.86	2.46	54.4	72.3	38.6–95.1
Sulphide			1.21	0.29–2.28					
Sulphate	500		102.5	101.05–122.10	68.58	48.20	114.34	218.12	10–252
Ammonia	0.08		47.34	41.48–95.16	0.86	78.68	86.4	122.1	83.9
Ammonium	2.0								0.1–7.0
Alkalinity					480.00	300.00	502	623	
Calcium	180		2570.00	2500.00–3751.40					30–182
Potassium	120		1800.00	1340.00–2250.00					
Sodium	50		79.20	126.50–7740.00					
Mg	40		2.40	4.00–12.00					18–175

Parameters	National criteria	Odo Oba (Osun State)	Aba Eku (Oyo State)		Olusosun (Lagos State)		Aba Eku	Olusosun	Aba Eku
			Raw	Simulated	Raw	Simulated			
Cu	0.01	0.0935	4.50	5.30–20.00	0.77	0.44	2.44	3.86	ND–0.103
Pb	0.1	0.0588	2.20	3.60–8.80	1.40	0.69	2.08	2.00	0.008–73.3
Fe	0.5	8.321			1.90	0.83	3.20	4.71	0.30–50.5
Cd	0.01	0.0385	2.20	3.65–8.82	0.58	0.46	1.44	2.20	0.4–5.7
Ag		0.0163							
Mn		0.253	2.80	4.20–15.00	0.79	0.46	2.90	3.10	0.6–23.8
Ni	0.1	0.249	3.60	1.80–5.12			1.88	2.51	ND–0.10
Ar			2.26	3.63–8.83					
Zn	0.2		3.60	5.00–18.00					0.3–3.5
Hg	0.0005		2.38	2.14–8.75	0.41	0.23			
Cr	0.5		2.28	2.50–8.70			2.32	2.43	0.04–2.5
As					0.36	0.27	1.50	2.60	
	[53]	[110]	[54]		[55]		[56]		[57, 58]

All in mg/l except pH and colour.

Table 9.
Leachate characteristics from dumpsites and landfills from Southwest Nigeria.

Parameters	National criteria	Effurun	Nnewi	Agbani	Abakaliki	Onitsha	Aba	Akoko Edo	Ibadan (Ona River)	
									2002	1997
		Delta State	Anambra State	Enugu	Ebonyi	Anambra	Abia	Ondo	Oyo State	
EC		628.0–694.5	140.3–197.0				18.6–790.2	43.4–48.60	366–611	160–600
TDS		80.3–694.5	12.28–16.82	10.0–30.0			10.3–855.8	320–364	408–2054	90–250
Total solids			16.81–21.7	40.0–380.0			80.0–81.7	535.0–600.0	460–2160	
Turbidity			180.0–338.54				0.04–32.1	3.62–5.91		
TSS	0.25							171–265	38–170	
Ca^{2+}	180	8.0–39.0		1.60–28.10			13.3–158.2		ND–4.0	
Mg^{2+}	40	12.0–214.0		ND–97.30			4.60–10.00			
Fe^{2+}	0.05	0.08–1.82		0.17–1.89		0.10–0.80	19.61–32.14			0.03–0.6
Na^{2+}	120	65.89–118.72							184–358	
K^{2+}	50	45.91–49.19							2.0–8.0	
Nitrate	9.1	0.87–1.25		0.22–2.43			0.20–8.20	8.04–8.28		
Phosphate	3.5	3.60–50.34	6.13–7.25				0.20–10.40	1.39–1.41	700.0–1129	
Sulphate	100	64.0–100.5	211.66–239.17				27.4–103.8	63.0–74.0	386–480	
Chloride	300	40.5–240.6	122.93–164.82	2.00–47.90			12.1–184.0	143–190	45.0–70.0	
Alkalinity			24.97–33.87				3.1–3.3			40.9–175.8
Acidity							2.0			
Ammonia-nitrogen										ND–2.2
Ammonium	0.05	1.02–3.24								
Total hardness				4.0–40.0			18.1–168.2		129–320	
%TOC							2.98–3.01			

Parameters	National criteria	Effurun	Nnewi	Agbani	Abakaliki	Onitsha	Aba	Akoko Edo	Ibadan (Ona River)	
									2002	1997
		Delta State	Anambra State	Enugu	Ebonyi	Anambra	Abia	Ondo	Oyo State	
Phenol									1.2–2.0	
pH	6.5–8.5	6.8–7.0	6.77–6.97	3.5–6.0			6.1–6.8	6.43–7.24		7.2–8.9
DO	>6.0		43.04–63.93	1.4–4.9			5.6–11.3			0.9 –22.1
BOD	3.0	6.8–8.9	12.67–20.55	5.0–18.0			3.8–37.9			0.00–11.4
COD	30.0	55.0–95.0	264.89–342.45	60.0–320.0			5.6–53.0			
Zinc	0.01		0.40–1.42		0.14–0.16	0.30–1.30	0.48–0.57		2.10–2.5	0.007–0.5
Aluminum	0.2								23–76	
Copper	0.001		0.40–0.08		0.9–1.0	010–0.90	0.001			ND
Chromium	0.001				0.02–0.04		0.001		3.0–4.0	ND–0.03
Cadmium	0.005				0.02–0.05	0.22–0.99	0.001		ND–2.50	ND–0.01
Iron	0.05		1.46–6.42						17–25	
Lead	0.01		0.23–0.31		0.06–0.08	0.11–1.99	0.06–0.09		32–51	ND–0.06
Nickel	0.01						0.053–0.06			
Arsenate	0.05					0.21–2.6				
Mercury	0.001					0.3–1.8				
Cobalt										0.02–0.2
Manganese										0.01–0.17
	[53]	[59]	[60]	[61]	[62]	[2]	[63, 64]	[65]	[66]	[67]

Table 10.
Quality of inland surface waters receiving dumpsite/landfill effluent/products from Nigerian cities.

Parameters	National criteria		Abuja	Ilokun	Effurun	Minna	Onitsha	Ota	Akoko Edo	Lagos	Ibadan	
										Olusosun/ Ojota	Ring road	Aba Eku
				Ekiti State	Delta State	Niger State	Anambra State	Ogun State	Ondo State	Lagos State	Oyo State	
Electrical conductivity (μS/cm)		1000.0	30–213		20.3–1200.0	344.0–1191.0		15–1572	13.6–51.8	107.0–4043.0	172–868	106.9–696.0
Total dissolved solids(mg/l)		500.0	65–132		9.7–765.4	210.0–738.9		8–836	102.0–415.0	40.0–2021.0	147–1100	53.9–347.0
Suspended solids (mg/l)			15–35						13.0–52.0		14.0–85.0	0.00–246.0
Total solids	1500.0								115.0–430.0	500.0–1370.0	160–1620	53.4–347.0
Turbidity (NTU)		5	1–9			4.5–38.7			1.2–2.3			
Phenol	2.0	0.001									0.20–1.0	
pH	6.5–9.2	6.5–8.5	6.8–7.2		6.3–7.1	7.2–8.4		4.5–6.01	5.8–7.0	3.8–7.0	5.56–8.22	7.4–8.3
Fe	1.0	0.3	ND–0.32	0.12–0.5	0.001–1.9		0.5–2.91	2.4–4.5		0.06–5.5	ND–21	ND–16.9
Mg	150.0	20.0		3.5–5.2	0.1–1.5		ND–18.72	1.5–13.1		4.6–74.9		1.6–84.9
Zn	15	3.0		0.2–0.4			0.18–0.60	2.0–3.1		1.01–2.7	0.9–3.6	ND–2.5
Mn	0.5	0.2		0.1–0.3			ND–0.55	0.7–0.9		0.03–1.3		ND–0.5
K					5.8–32.2			2.1–2.9		1.0–4.1	0.6–2.5	0.9–52.4
Na		200.0			10.7–65.4			5.8–7.1		4.0–13.4	104–292	
Ca	200			29.1–72.1	11.3–38.0	71.0–327.0		4.0–89.9		4.0–98.2	1.0–9.0	3.7–87.5

Parameters	National criteria		Abuja	Ilokun	Effurun	Minna	Onitsha	Ota	Akoko Edo	Lagos	Ibadan	
										Olusosun/ Ojota	Ring road	Aba Eku
				Ekiti State	Delta State	Niger State	Anambra State	Ogun State	Ondo State	Lagos State	Oyo State	
Cd	0.006	0.003		ND–0.001			0.22–0.24			0.004–0.007	ND–3.6	0.01–0.2
Ni	0.075			ND–0.007			ND–0.03	0.01–0.03		0.02–0.03		ND–0.22
Cr	0.03	0.05		0.007–0.01			ND–0.002	0.01–0.03		0.009–0.12	1.0–6.0	ND–0.05
Cu	0.075	1.0		0.02–0.4			0.03–0.2	0.29–0.67		0.04–0.6		ND–0.04
Pb	0.075	0.01		0.001–0.03			0.19–0.5	ND–0.03		0.003–0.08	ND–58	ND–0.2
As	0.06	0.01		0.001–0.001			0.003–0.5					
Silver							ND–0.02					
Aluminum		0.2					ND–0.007				19–42	
Molybdenum							ND–0.9					
Mercury	0.0003	0.001					0.002–0.4					
Cobalt	0.1						ND–0.081			0.025–0.001		
Cl	600	250.0		39.6–216.7	8.9–225.0	28.1–167.9		70.9–186.5	126.0–304.0	53.1–726.0	20.00–118.00	1.5–68.9
Sulphate	400	100.0	20–231	2.6–6.2	ND–24.3			11–278	54.0–130.0	2.0–735.0	114–700	1.6–43.2
Nitrate (N)	50	50.0	3.6–8.0		0.08–56.0			1.3–16.7	4.4–8.8	ND–45.0		0.1–44.2
NO_3			0.2–41.5	0.02–0.2						9.3–66.0		

Parameters	National criteria		Abuja	Ilokun	Effurun	Minna	Onitsha	Ota	Akoko Edo	Lagos	Ibadan	
										Olusosun/ Ojota	Ring road	Aba Eku
				Ekiti State	Delta State	Niger State	Anambra State	Ogun State	Ondo State	Lagos State	Oyo State	
Nitrite	2	0.2		ND		0.001–0.2				0.06–0.98		
Ammonium					ND–0.9					0.16–96.0		0.03–0.7
Total hardness			19.4–79.0			112.0–444.0		10–212		45.0–367.0	78.8–428	
Hardness (Ca) $CaCO_3$	500	150.0	28.7–48.7									
Hardness (Mg) $CaCO_3$			28.7–48.7									
Total alkalinity (mg/l)			4.00–74.0									
Phosphate as phosphorus			0.01–0.2	ND–0.3	ND–8.3	0.3–0.9		10.3–42.	0.4–0.9	0.02–40.8	257–1040	
BOD					ND–16.4	4.1–8.1				40.0–3427.0		
COD					ND–35.0					1.8–3170.0		4.2–18.7
DO						3.6–9.5				0.5–0.7		
	[53]	[68]	[69]	[70]	[59]	[71]	[2, 72]	[73]	[64]	[72, 74, 75]	[66]	[76]

Table 11.
Quality of inland groundwaters receiving dumpsite/landfill effluent/products from Nigerian cities.

chromium, copper, lead, arsenic and aluminum and cobalt in groundwater from most locations should be a cause for concern and perhaps necessitates detailed nationwide surveillance, considering the proportion of population dependent on groundwater. The intake of these metals has been implicated in a variety of human ailments leading to severe problems via disruption of metabolic functions in two ways [81]:

1. They accumulate and thereby disrupt function in vital organs and glands such as the heart, brain, kidneys, bone, liver, etc.

2. They displace the vital nutritional minerals from their original place, thereby hindering their biological function.

Residents around the dumpsites are partly or wholly dependent largely on either surface or groundwater for direct or indirect daily water requirements. Thus contact with these water bodies is inevitable, even at distances considered areas with no likely effects. Determination of the health implications of such contacts at present has not been clearly defined, from very limited reports on public health aspects of dumpsite managements. This is because it has not been possible to separate consequences of dumpsite contaminated surface and groundwater contacts from medical conditions associated with population living around dumpsites. Studies [82–85] reported the following: inhalation of odor, exposure to dust, exposure to smoke, exposure through water sources, consumption of plant materials, consumption of animal materials, exposure through organisms (vectors), noise from vehicles, exposure to fire, dermal contacts and exposure through domestic animals as possible routes of human exposure and contact with dumpsites and products of dumpsite modifications. Medical conditions reported from the population living close to dumpsites in different parts of Nigeria are presented in **Table 12**, which have been observed in Nigeria from areas of regular contacts with contaminated water [90] but not from dumpsites or landfills. The implication of the above is that symptoms may indicate conditions from multiple exposures or contacts. Inland waters in Nigeria have been subjected to inundations with inputs from multiple sources with resultant changes in quality. The almost hidden nature of contamination and contamination routes by dumpsites reinforces the dangers of not paying required attention to dumpsites, associated activities and value chain. This is because each step or link has an effect on inland water and hence human population making these sources of contamination very dangerous and harmful. Therefore, numerous health hazards associated with waste dump sites in major economic centres in Nigeria [27, 91, 92] can be said to be largely denominated by the resident and/or dominant waste components.

Radionuclides have also been reported and associated with dumpsites and landfills in Lagos State [93–95], Oyo State [95–97], Ogun State [98–100], Plateau State [101], Benue State [101], Ekiti State [95], Rivers State [102–105] and Delta State [106]. These dangerous natural and artificial radiation materials from unregulated and unmanaged dumpsites and landfills released into inland water sources pose risks to resident organisms and population of humans, dependent directly on water for domestic purpose and consumption of resident aquatic organisms.

Radionuclides have been reported in leachates [102] and groundwater [102, 105, 107] and rivers [107, 108] with identified sources being the human activities, inclusive of dumpsites [102, 105] and abattoir wastes [109]. Dumpsites and landfills are therefore potential sources of radionuclide inputs into inland surface

Lagos	Port Harcourt/Owerri/Aba
Asthma	High temperature and fever/typhoid
Bronchitis	Watery stool/frequent stooling
Chest pain	Vomiting
Lung disease	Catarrh and cough
Nose/throat problems	Loss of appetite
Breathing	Pains in the abdomen and body
Tuberculosis	Dizziness
Skin infection	Blood spotted stool
Headaches/nausea/diarrhea/dysentery	Urinary tract infection
Children's diseases	Acute osteomyelitis
Accident/injury	
Malaria	
[84, 86, 87]	[88, 89]

Table 12.
Ailments associated with population living near dumpsites.

and groundwaters; the above-cited reports indicated the presence of radionuclides in soils around target dumpsites, confirming the migration of substances from dumpsites, as reported [108], using time-lapsed vertical electrical sounding (VES). This migration of materials into ground- and surface waters will facilitate exposure of resident and non-resident population to radioactive material by direct or indirect intake, respectively. Low cancer risks from chronic exposure to radiation from dumpsites in Nigeria have been suggested [97] even at the low level, thus further establishing the need for urgent management strategies for MSW in Nigeria.

4. Conclusion and recommendation

Nigeria's development is currently enjoying active support of multilateral agencies, with the sole aim of expanding and diversifying the economy through, but not limited to, multinational manufacturing and small- and medium-scale enterprises. These are desirable and needed to improve socioeconomic status of the populace. However, complementary in-depth consideration of the ecological consequences of expanded economy must include increased generation of MSW, which usually begin with unregulated and undocumented dumpsites associated with penetration of economic activities. The inability of agencies responsible for waste management to anticipate and plan for the increase of MSW is the major reason for the surge in MSW generation and persistence. These will eventually become sources of sometime unexplained inland water contamination and/or public health problems or outbreaks. In view of this, the following be deeply considered to minimize the negative impacts of MSW on inland waters:

1. Collection of dumpsite and landfill history and location data in each local government area (LGA) nationwide.

2. Characterize wastes associated with each dumpsite and landfill, to provide data for risk assessment of dumpsite or landfill products.

3. Information on nearby surface and groundwater and their utilization by residents.

4. Information on geophysical assessment of pollutant movements in soil.

5. Regular determination of inland water quality in the vicinity.

6. Create awareness on the need to sort waste from source before disposal.

7. Encourage adoption of recycle and reuse of wastes to reduce wastes generated.

8. Undertake spatial analyses of population or residents' socioeconomic characteristics to predict waste profiles and determine appropriate management MSW strategy.

Conflict of interest

I declared no conflict of interest.

Author details

Akindayo A. Sowunmi
Department of Zoology, Hydrobiology and Fisheries Unit, University of Ibadan, Ibadan, Nigeria

*Address all correspondence to: dayolegba@gmail.com

References

[1] World Bank Group. Available from: https://data.worldbank.org/country/nigeria [Accessed: November 26, 2018]

[2] Ezeabasili ACC, Anike OL, Okoro BU. Urban water pollution by heavy metals and the implication in Onitsha, Nigeria. African Journal of Environmental Science and Technology. 2015;**9**(4):325, 331

[3] Ita EM, Sado EK. A preliminary inventory of water bodies in Nigeria with special reference to ponds, lakes and reservoirs. In: Proceedings of 3rd Annual Conference of the Fisheries Society of Nigeria (F ISON) Maiduguri, 22-25 February; 1983

[4] Ita EO, Sado EK, Balogun JK, Pandogari A, Ibitoye B. Inventory survey of Nigeria inland waters and their reference to ponds, lakes, reservoirs and major rivers. Kainji Lake Research Institute, Technical Report Series. 1985;**14**:51

[5] Ayoade JO. Water resources and their development in Nigeria. Hydrological Sciences Journal. 1975;**20**(4):581-591. DOI: 10.1080/02626667509491589

[6] Ezechi EH, Nwabuko CG, Enyinnaya OC. Municipal solid waste management in Aba, Nigeria: Challenges and prospects. Environmental Engineering Research. 2017;**22**(3):231-236. DOI: 10.4491/eer.2017.100

[7] Worldometers. Dover, Delaware, U.S.A. Available from: http://www.worldometers.info/world-population/nigeria-population/ [Retrieved: November 26, 2018]

[8] Achi HA, Adeofun CO, Gbadebo AM, Ufoegbune GC, Oyedepo JA. An assessment of solid waste management practices in Abeokuta, Southwest, Nigeria. Journal of Biological and Chemical Research. 2012;**29**(2):177-188

[9] Babayemi JO, Dauda KT. Evaluation of solid waste generation, categories and disposal options in developing countries: A case study of Nigeria. Journal of Applied Sciences and Environmental Management. 2009;**13**(3):83-88

[10] Ayuba KA, Manaf LA, Sabrina AH, Nur Azmin SW. Current status of municipal solid waste management practise in FCT Abuja. Environmental and Earth Sciences Research Journal. 2013;**5**(6):295-304

[11] Babatunde BB, Vincent-Akpu IF, Woke GN, Atarhinyo E, Aharanwa UC, Green AF, et al. Comparative analysis of municipal solid waste (MSW) composition in three local government areas in Rivers State, Nigeria. African Journal of Environmental Science and Technology. 2013;**7**(9):874-881

[12] Oyelola O, Ajiboshin I, Okewole J. Effective solid waste management in the circular economy: The case study of Lagos state, Nigeria. Procedia Environmental Science, Engineering and Management. 2017;**4**(3):143-148

[13] Nwaokobia K, Ogboru RO, Okolie PL. Solid waste management: Efficient approach towards sustainable development in Nigeria. Greener Journal of Environmental Management and Public Safety. 2013;**7**(2):035-042. DOI: 10.15580/GJEMPS.2018.2.090117118

[14] Odigie JO. Harmful effects of wastewater disposal into water bodies: A case review of the Ikpoba River, Benin City, Nigeria. Tropical Freshwater Biology. 2014;**23**:87-101

[15] Ajero CMU, Chigbo UN. A study on the evaluation of industrial solid waste management approaches in some industries in Aba, South Eastern Nigeria. West African Journal of

Industrial and Academic Research. 2012;**4**(1):103-112

[16] Kadafa AA. Solid waste management practice of residents in Abuja municipalities (Nigeria). IOSR Journal of Environmental Science, Toxicology and Food Technology. 2017;**11**(2):87-106

[17] Abdullah TPL, Bena R. Solid waste management practices in Kebbi State, Nigeria: Problems and prospect. International Refereed Journal of Engineering and Science. 2013;**2**(12):48-54

[18] Limana A, Ngahb I. Appraisal of municipal solid waste management (MSWM) in Jimeta-Yola Adamawa State of Nigeria: The need for innovative, PPP strategies. Jurnal Teknologi. 2015;**77**(15):93-102

[19] Efe SI. Waste disposal problems and management in Ughelli, Nigeria. Journal of Environmental Protection. 2013;**4**:4-11

[20] Olabode AD, Akinola L. Environmental impact of indiscriminate waste disposal on river channel in part of Akoko-region, Ondo State, Nigeria. International Journal of Innovation and Scientific Research. 2014;**5**(2):162-168

[21] Sha'Ato R, Aboho SY, Oketunde FO, Eneji IS, Unazi G, Agwa S. Survey of solid waste generation and composition in a rapidly growing urban area in Central Nigeria. Waste Management. 2007;**27**(3):352-358

[22] Abur BT, Oguche EE, Duvuna GA. Characterization of municipal solid waste in the Federal Capital Abuja, Nigeria. Global Journal of Science Frontier Research: Environment and Earth Science. 2014;**14**(2):1-7

[23] Dauda M, Osita OO. Solid waste management and re-use in Maiduguri, Nigeria. In: Harvey, P, editor. Towards the Millennium Development Goals—Actions for Water and Environmental Sanitation: Proceedings of the 29th WEDC International Conference; 22-26 September; Abuja, Nigeria; 2003. pp. 20-23

[24] Buba HB. Assessment of household solid waste management in Gombe, Nigeria [MSc thesis]. Zaria, Nigeria: Department of Urban and Regional Planning, Ahmadu Bello University; 2016

[25] Imam A, Mohammed B, Wilson DC, Cheeseman CR. Solid waste management Abuja, Nigeria. Waste Management. 2008;**28**:468-472

[26] Ngwuluka N, Ochekpe N, Odumosu P, John SA. Waste management in healthcare establishments within Jos Metropolis, Nigeria. African Journal of Environmental Science and Technology. 2009;**3**(12):459-465

[27] Igbinomwanhia DI. Characterization of commercial solid waste in Benin Metropolis, Nigeria. Journal of Emerging Trends in Engineering and Applied Sciences. 2012;**3**(5):834-838

[28] Sridhar MKC, Bammeke AO, Omisakin MA. A study on the characteristics of refuse in Ibadan, Nigeria. Waste Management and Research. 1985;**3**(3):191-201

[29] Bammeke AO, Sridhar MKC. Market wastes in Ibadan, Nigeria. Waste Management and Research. 1989;**3**(3):191-201

[30] Ajadi BS, Tunde AM. Spatial variation in solid waste composition and management in Ilorin Metropolis, Nigeria. Journal of Human Ecology. 2010;**32**(2):101-108

[31] Nabegu AB. An analysis of municipal solid waste in Kano

Metropolis, Nigeria. Journal of Human Ecology. 2010;**31**(2):111-119

[32] Asibor IG, Edjere O. Assessment of the activities of scavengers and their economic impacts in waste recovery in Warri Metropolis, Delta State Nigeria. International Research Journal of Public and Environmental Health. 2017;**4**(2):22-29

[33] Ahmed R, Onyidoh HE. Municipal solid wastes (MSW) management in high population density areas of Zaria and Kaduna Metropolis in Kaduna State, Nigeria. Greener Journal of Environmental Management and Public Safety. 2017;**7**(2):023-034

[34] Nwachukwu MU. Solid waste generation and disposal in a Nigerian city: An empirical analysis in Onitsha Metropolis. Journal of Environmental Management and Safety. 2010;**1**(1):180-191

[35] Angaye TC, Daokoru-Olukole C, Abowei JN. Characterization of household solid wastes in the Niger Delta: A case study of 100-domestic units in Yenagoa Metropolis, Nigeria. International Journal of Sustainable Development Research. 2017;**3**(5):50-53. DOI: 10.11648/j.ijsdr.20170305.12

[36] Belel ZA, Mahmoud H. Survey of municipal solid waste in Jimeta-Yola, Northeastern Nigeria. International Journal of Scientific and Engineering Research. 2013;**4**(1):1-4

[37] Egbere JO, Itelima JU, Opiah FO. Municipal solid wastes segregation and their health hazard implication in Angwan-Rogo Ward of Jos, Plateau State, Nigeria. Journal of Environmental Sciences. 2001;**4**(1):1-4

[38] Akpen GD, Aondoakaa SC. Assessment of solid waste management in Gboko Town. Global Journal of Environmental Sciences. 2009;**8**(2):71-77

[39] Roberts CL, Watkin GD, Ezeah C. Seasonal variation and municipal solid waste composition—Issues for development of new waste management strategies in Abuja, Nigeria. Journal of Solid Waste Technology and Management. 2010;**36**(4):210-219

[40] Kofoworola OF. Recovery and recycling practices in municipal solid waste management in Lagos, Nigeria. Waste Management. 2007;**27**:1139-1143

[41] Oyelola OT, Babatunde AI. Characterization of domestic and market solid wastes at source in Lagos Metropolis, Lagos, Nigeria. African Journal of Environmental Science and Technology. 2008;**3**(12):430-437

[42] Lagos State Waste Management Authority (LAWMA). Solid Waste Data in the City of Lagos. Department of Planning Research and Statistics. 2011. Available from: http://www.lawma.gov.ng/lawma_prs.html [Accessed: October 16, 2018]

[43] Lagos State Waste Management Authority (LAWMA). Solid Waste Data in the City of Lagos. Department of Planning Research and Statistics. 2012. Available from: http://www.lawma.gov.ng/lawma_prs.html [Accessed: October 16, 2018]

[44] Audu HAP, Aigwi IE, Enaboifo MA. Solid waste composition analysis for the development of a suitable waste disposal system in Port Harcourt L.G.A of Rivers State, Nigeria. Journal of Emerging Trends in Engineering and Applied Sciences. 2015;**6**(2):113-119

[45] Akor AJ, Ayotamuno MJ, Aman LI, Enokela SO. Assessment of domestic solid waste generation in Port Harcourt by separator-receptacle technology. International Journal of Scientific and Engineering Research. 2017;**4**(1):1-7

[46] Ogbonna DN. The impact of untreated sewage wastes discharge

on the physico-chemical properties of rivers in Port Harcourt Metropolis. World Journal of Scientific Research and Reviews. 2014;**2**(2):1-19

[47] Binafeigha TR, Enwin A. The state of solid waste management in Port Harcourt City, Nigeria. American Journal of Civil Engineering and Architecture. 2017;**5**(4):160-166. DOI: 10.12691/ajcea-5-4-4

[48] Ukem EO. Electronic waste: A growing challenge in Nigeria. Global Journal of Pure and Applied Sciences. 2018;**14**(4):459-462

[49] Nnorom IC, Osibanjo OO. Electronic waste (e-waste): Material flows and management practices in Nigeria. Waste Management. 2008;**28**(8):1472-1479

[50] Ajayi SO, Adelaye SA. Pollution studies on Nigerian rivers, I: Preliminary report on the pollution level of River Ona and River Ogunpa. Bulletin of the Chemical Society of Nigeria. 1977;**2**:71-86

[51] Ajayi SO, Osibanjo OO. Pollution studies on Nigerian rivers, II: Water quality of some Nigerian rivers. Environmental Pollution (Series B). 1981;**2**:87-95

[52] Pohland FG, Harper SR. Critical Review and Summary of Leachate and Gas Production from Landfills. Cincinnati, OH, U.S.A.: U.S. Environmental Protection Agency; 1986. EPA/600/2-86/073

[53] National Environmental Standards and Regulations Enforcement Agency (NESREA). National Environmental (surface and groundwater quality) Regulations. Federal Republic of Nigeria (FGN) - Official Gaezette. 2011;**98**(49):B693-B727

[54] Bakare AA, Wale-Adeyemo AR. The mutagenic and cytotoxic effects of leachates from domestic solid wastes and Aba-Eku landfill, Nigeria on *Alliums cepa*. Nature, Environment and Pollution Technology. 2004;**3**(4):455-462

[55] Alimba CG, Bakare AA, Aina OO. Liver and kidney dysfunction in Wistar rats exposed to municipal landfill leachate. Resources and Environment. 2012;**2**(4):150-163. DOI: 10.5923/j.re.20120204.04

[56] Alimba CG, Bakare AA, Arinola GO. Effects of municipal solid waste leachate on leucocyte and differential counts in rat. Global Journal of Environmental Research. 2009;**3**(3):205-210

[57] Oshode OA, Bakare AA, Adeogun AO, Efuntoye MO, Sowunmi AA. Ecotoxicological assessment using *Clarias gariepinus* and microbial characterization of leachate from municipal solid waste landfill. The International Journal of Environmental Research. 2008;**2**(4):215-225

[58] Oni AA, Hassan AT, Peijun L. Toxicity of Leachates from the Aba-Eku Landfill Leachate Lagoon, Ibadan, South-Western Nigeria. Advances in Applied Science Research. 2011;**2**(2):450-460

[59] Ohwoghere-Asuma O, Aweto KE. Leachate characterization and assessment of groundwater and surface water qualities near municipal solid waste dump site in Effurun, Delta State, Nigeria. Journal of Environment and Earth Science. 2013;**3**(9):126-134

[60] Okeke IE, Ekenta EO, Mbanefo D. Effects of solid wastes on Ukwaka Stream, Nnewi, Nigeria. International Journal of Advances in Scientific Research and Engineering. 2018;**4**(5):172-185

[61] Nwodo PO. The effects of poor wastes disposal on the stream water

quality in Agbani Urban Area in the Nkanu West LGA of Enugu State. [PGD]. Nsukka, Nigeria: Department of Geography, University of Nigeria; 2012

[62] Njoku C. Effect of waste dumpsites on water and air qualities in Abakaliki, Southeastern Nigeria. International Journal of Plant and Soil Science. 2015;**4**(5):455-460

[63] Nkwocha EE, Pat-Mbano EC, Nnaji AO. Effect of solid waste dump on river water quality: A paradigm in a Nigerian tropical environment. International Journal of Science and Nature. 2011;**2**(3):501-507

[64] Iwuoha GN, Osuji LC, Onwuachu UI. Impact of waste dump on the sediment and surface water quality of Otamiri river, Nigeria. Journal of Applied Sciences and Environmental Management. 2013;**17**(4):573-575

[65] Ololade IA, Adewunmi A, Ologundudu A, Adeleye A. Effects of household wastes on surface and underground waters. International Journal of Physical Sciences. 2009;**4**(1):022-029

[66] Odukoya OO, Arowolo TA, Bambose O. Effects of solid waste landfill on underground and surface water quality at Ring Road, Ibadan, Nigeria. Global Journal of Environmental Sciences. 2002;**1**(1):43-52

[67] Oduwole GA. Indices of pollution in Ogunpa and Ona rivers, Ibadan: Physico-chemical, trace metal and plankton studies [PhD thesis]. Nigeria: Department of Zoology, University of Ibadan; 1997

[68] Standard Organization of Nigeria. Nigerian Standard for Drinking Water Quality. NIS-554-2015

[69] Musa JJ. Effect of domestic waste leachates on quality parameters of groundwater. Leonardo Journal of Sciences. 2014;**24**:28-38

[70] Olagunju E, Badmus O, Ogunlana F, Babalola M. Environmental impact assessment of waste dumpsite using integrated geochemical and physico-chemical approach: A case study of Ilokun waste dumpsite, Ado - Ekiti, Southern Nigeria. Civil Engineering Research Journal. 2018;**4**(2):001-0013. DOI: 10.19080/CERJ.2018.04.555631

[71] Saidu M. Effect of refuse dumps on ground water quality. Advances in Applied Science Research. 2011;**2**(6):595-599

[72] Michael EI. Evaluation of leachate contamination index obtained from dumpsite in Onitsha, Nigeria. Journal of Environmental & Analytical Toxicology. 2017;**7**(6):525. DOI: 10.4172/2161-0525.1000525

[73] Ameloko AA, Ayolabi EA, Enaworu E, Bolujo E. Assessment of leachate contamination of groundwater around the Igbenre Ekotedo dumpsite, Ota, Southwest Nigeria. Petroleum and Coal. 2018;**60**(5):890-902

[74] Adeyemi O, Oloyede OB, Oladiji AT. Physicochemical and microbial characteristics of leachate-contaminated groundwater. Asian Journal of Biochemistry. 2007;**2**(5):343-348

[75] Sanusi AL. Groundwater quality assessment at Olusosun landfill, Lagos, Nigeria [MSc thesis]. British Columbia: Royal Roads University Victoria; 2013

[76] Oni AA, Hassan AT. Groundwater quality in the vicinity of Aba-Eku dumpsite, Ibadan, SW Nigeria: A detailed report. Ethiopian Journal of Environmental Studies and Management. 2013;**6**(6):589-600. DOI: 10.4314/ejesm.v6i6.2

[77] Alimba CG, Bakare AA. In-vivo micronucleus test in the assessment of

cytogenotoxicity of landfill leachates in three animal models from various ecological habitats. Ecotoxicology. 2016;**25**(2):310-319. DOI: 10.1007/s10646-015-1589-3

[78] Alimba CG, Saliu JK, Adesanya A, Bakare AA. Evaluation of genotoxicity of a municipal landfill leachate by micronucleus test using *Clarias gariepinus*. Research in Environment and Life Sciences. 2011;**4**(1):1-6

[79] Ahianba JE, Dimuna KO, Okogun GRA. Built environment decay and urban health in Nigeria. Journal of Human Ecology. 2008;**23**(3):259-265

[80] Nwankwoala HO. Groundwater and poverty reduction: Challenges and opportunities for sustainable development in Nigeria. Hydrology Current Research. 2016;**7**(2):240. DOI: 10.4172/2157-7587.1000240

[81] Singh R, Gautam N, Mishra A, Gupta R. Heavy metals and living systems: An overview. Indian Journal of Pharmacology. 2011;**43**(3):246-253

[82] Nduka JKC, Orisakwe OE, Ezenweke LO, Chendo MN, Ezenwa TE. Heavy metal contamination of foods by refuse dump sites in Awka, Southeastern Nigeria. The Scientific World Journal. 2008;**8**:941-948

[83] Magaji JY. Effects of waste dump on the quality of plants cultivated around Mpape dumpsite FCT Abuja, Nigeria. Ethiopian Journal of Environmental Studies and Management. 2012;**5**(4):567-573

[84] Alimba CG. DNA and systemic damage induced by landfill leachates and health impacts of human exposure to landfills in Lagos and Ibadan, Nigeria [PhD thesis]. Nigeria: Department of Zoology, University of Ibadan; 2013

[85] Obasi NA, Obasi SE, Aloh GO, Elom SO. Review of edible plants in dumpsites: Risks of heavy metals toxicity and implications for public health. Journal of Experimental Agriculture International. 2017;**7**(6):1-11

[86] Olorunfemi FB. Living with waste: Major sources of worries and concerns about landfills in Lagos Metropolis, Nigeria. Ethiopian Journal of Environmental Studies and Management. 2009;**2**(2):12-19

[87] Babs-Shomoye F, Kabir R. Health effects of solid waste disposal at a dumpsite on the surrounding human settlements. Journal of Public Health in Developing Countries. 2016;**2**(3):268-227

[88] Onwubiko JA. A study of the health effects of open refuse dumps on nearby residents in Eastern Nigeria. Sky Journal of Soil Science and Environmental Management. 2017;**6**(2):26-30

[89] Wachukwu CK, Mbata CA, Nyenke CU. The health profile of waste scavengers (rag pickers) in Port Harcourt, Nigeria. Journal of Applied Sciences. 2010;**10**(17):1968-1972

[90] Karshima SO. Public health implications of poor municipal waste management in Nigeria. Vom Journal of Veterinary Science. 2016;**11**:142-148

[91] Ogbonna DN, Ekweozor IKE, Igwe FU. Waste management: A tool for environmental protection in Nigeria. AMBIO: A Journal of the Human Environment. 2002;**31**(1):55-57

[92] Alam P, Ahmade K. Impact of solid waste on health and the environment. Journal of Sustainable Development and Green Economics. 2013;**2**(1):2

[93] Oladapo OO, Oni EA, Olawoyin AA, Akerele OO, Tijani SA. Assessment of natural radionuclides level in wasteland soils around Olusosun Dumpsite Lagos,

Nigeria. IOSR Journal of Applied Physics. 2012;**2**(3):38-43

[94] Olubosede O, Akinnagbe OB, Adekoya O. Assessment of radiation emission from waste dumpsites in Lagos State of Nigeria. International Journal of Computational Engineering Research. 2012;**2**(3):806-811

[95] Jibiri NN, Isinkaye MO, Momoh HA. Assessment of radiation exposure levels at Alaba e-waste dumpsite in comparison with municipal waste dumpsites in southwest Nigeria. Journal of Radiation Research and Applied Science. 2014;7(4)536-541. DOI: 10.1016/j.jrras.2014.09.002

[96] Odunaike RK, Laoye JA, Alausa SK, Ijeoma GC, Adelaja AD. Radiation emission characterization of waste dumpsites in the city of Ibadan in Oyo State of Nigeria. Research Journal of Environmental Toxicology. 2008;**2**:100-103

[97] Isola GA, Oni MO, Olasunkanmi SK.Radionuclide concentrations in soil around waste dumpsites and its excess lifetime cancer risk due to gamma radioactivity in Ogbomoso Metropolis, South-western Nigeria. BEST: International Journal of Humanities, Arts, Medicine and Sciences. 2015;**3**(6):25-30

[98] Odunaike RK, Alausa SK, Fasunwon OO, Orunsolu BA, Ijeoma GC. Radiation dose survey of refuse dumpsites in Abeokuta in Ogun State in Southwest zone of Nigeria. Research Journal of Environmental Sciences. 2008;**3**:262-266

[99] Ademola AK, Babalola AI, Onyinye FO, Onuh AD, Enyenihi EE. Assessments of natural radioactivity and determination of heavy metals in soil around industrial dumpsites in Sango-Ota, Ogun state, Nigeria. International Journal of Medical Physiology. 2014;**39**:106-111

[100] Usikalu MR, Ola OO, Achuka JA, Babarimisa IO, Ayara WA. Background radiation dose of dumpsites in Ota and Environs. International Conference on Space Science and Communication. IOP Conference Series: Journal of Physics. 2017;**852**:012045. DOI: 10.1088/1742-6596/852/1/012045

[101] Sombo T, Avaa AA, Eweh EJ, Akohu M, Akine A, Shaibu B, et al. Radioactivity of dumpsites within selected towns of Benue and Plateau State, North Central Nigeria. IOSR Journal of Applied Physics. 2018;**10**(3):15-20

[102] Ehirim CN, Itota GO. Radiological impact of a municipal solid waste dumpsite on soil and groundwater using 2-D resistivity tomography and gamma ray spectroscopy. IOSR Journal of Environmental Science, Toxicology and Food Technology. 2013;**2**(5):35-42

[103] Avwiri GO, Olatubosun SA. Assessment of environmental radioactivity in selected dumpsites in Port Harcourt, Rivers State, Nigeria. International Journal of Scientific and Technology Research. 2014;**3**(4):263-269

[104] Ulakpa WC, Eyankware EO, Eyankware MO, Eyankware ROU. Evaluation of radionuclides in Eliozu dumpsite, Obio-Akpor L.G.A. South-South Nigeria. International Journal of Science and Healthcare Research. 2016;**1**(2):15-22

[105] Itota GO, Balogun AO. The radiological impact of a municipal solid waste dumpsite on soil and groundwater using gamma ray spectroscopy. Physical Science International Journal. 2017;**13**(4):1-6

[106] Avwiri GO, Esi EO. Evaluation of background ionization radiation level in some selected dumpsites in Delta State, Nigeria. Advances in Physics Theories and Applications. 2014;**35**:36-42

[107] Agbalagba EO, Avwiri GO, Chad-Umoren YE. Gross α and β activity concentration and estimation of adults and infants dose intake in surface and ground water of ten oil fields environment in Western Niger Delta of Nigeria. Journal of Applied Sciences and Environmental Management. 2013;**17**(2):267-277

[108] Ameloko AA, Ayolabi EA. Geophysical assessment for vertical leachate migration profile and physicochemical study of groundwater around the Olusosun dumpsite Lagos, South-West Nigeria. Applied Water Science. 2018;**8**:142. DOI: 10.1007/s13201-018-0775-x

[109] Jibiri NN, Isinkaye MO, Ilori OA, Sowunmi AA. Determination of natural radionuclides in abattoir wastes and sediments from rivers and streams around selected abattoirs in Southwest Nigeria. International Journal of Low Radiation. 2015;**10**(2):93-106

[110] Bakare AA. The potential mutagenic effect of the leachates from rural solid waste landfill on *Allium cepa (L)*. SINET: Ethiopian Journal of Science. 2001;**24**(2):283-291

Section 3

Inteligent Techniques for Controling Municipal Solid Wastes

Chapter 6

Life Cycle Inventory (LCI) Modeling of Municipal Solid Waste (MSW) Management Systems in Kosodrza, Community of Ostrów, Poland: A Case Study

Dariusz Sala and Bogusław Bieda

Abstract

The purpose of this study is to perform the life cycle assessment (LCA) limited to life cycle inventory (LCI) related to municipal solid waste operating in Kosodrza, community of Ostrów, in Poland. The current LCI is a representative for year 2015 by application of PN-EN ISO 14040. The system boundary was labeled as gate-to-gate. The data used in this study, involving consumption of energy and fuels, water, materials, and waste, is obtained from (i) site-specific measured or calculated data and (ii) secondary data taken from integrated permit issued by Marshal of the Podkarpackie region in Rzeszów for the establishment of municipal services in Ostrów by entering the records concerning the waste landfill in Kosodrza. This study is based on the deterministic approach to LCI. Hence, uncertainty analysis is not carried out. The LCI model can be used in full LCA study.

Keywords: Poland, life cycle inventory, life cycle assessment, municipal solid waste management, landfill

1. Introduction

The traditional consideration of waste as a pollution has progressively shifted toward a new perspective, in which waste is regarded as a resource that could support societies to become more sustainable [1].

LCA as a tool to analyze waste management systems appeared in the early 1990s.

A number of models for LCA of waste management have been developed, and some of these models are commercially available, while others are affordable only to researchers [2]. All models are developed within the framework of LCA of waste management, and most models also include some kind of economic accounting [2]. Moreover, in [2], a summary of key features of waste management LCA models is presented. Among them are:

- *Integrated waste management-2 (IWM-2)*, updated version of the IWM-1, released by Procter and Gamble in 1995. The IWM-2 and IWM-1 have been used in many case studies in Europe, North and South America and Australia.

- *IWM Canada* was developed in Canada in partnership between Environment Canada and two industry associations. The model is built on an Excel platform and runs with a Visual Basic interface. The model has been used by municipalities across Canada by more than 250 registered users in evaluating environmental and economic impacts of existing or planned waste management systems. In the City of London, the model has been used in the implementation of a continuous improvement system for waste management, and it is used by universities.

- *ORWARE* model (*or*ganic *wa*ste *re*search) was developed in collaboration with several Swedish research institutes and universities. ORWARE was first developed as a tool for systems analysis of organic waste management. ORWARE is implemented in Matlab and in Excel. Several projects have been commissioned by Swedish municipalities, and also it is used in education at universities.

- *Solid waste management (ISWM, MSW-DST)*. The MSW-DST was designed to explore and evaluate the environmental aspect and cost of integrated MSW strategies. The model has been applied in local and regional MSW planning and evaluation for cities, counties and states across the Unites States. This model has also been used by the US Navy to develop an improved waste management plan that meets environmental targets at reduced cost.

- *WISARD* was first developed in 1999 by Ecobilan on behalf of Eco-Emballages in France and the Environment Agency of England and Wales. WISARD has been used by more than 50 local authorities and others in the United Kingdom in the development of regional and MSW strategies, and it has also been used in the development of the Scottish National Waste Plan.

- *Municipal solid waste management system assessment tool LCA-IWM* is a result of a project funded by the European Fifth Framework Program and consists of decision support tools: the waste prognostic tool and the municipal solid waste management system (MSWMS) assessment tool. MSWMS has been applied in case studies of different cities in fast-growing regions in Europe; some examples are Xanthi (Greece), Kaunas (Lithuania), Wrocław (Poland), Nitra (Slovakia) and Reus (Spain).

- *Environmental assessment of solid waste systems and technologies (EASEWASTE)* was developed by researchers at the Technical University of Denmark. Detailed scope of EASEWASTE is presented in [2]. EASEWASTE is designed to compare different waste management strategies, waste treatment methods and waste process technologies and to identify significant sources of environmental problems of the system.

- *Waste and resources assessment tool for the environment (WRATE)* was designed to address environmental aspects and impacts of municipal solid waste management (MSWM), and it was developed on behalf of the Environment Agency for England and Wales, Scottish Environment Protection Agency and Department of Environment. The default database includes 160 waste management technology datasets and energy mix for 40 countries (average and marginal) over a 20-year forecast. Moreover, WRATE also includes a database on materials and their inventories (Ecoinvent database), a default waste composition (national UK waste composition) and the most used impact assessment methods. Detailed scope in terms of material and energy flows and processes is illustrated in [2].

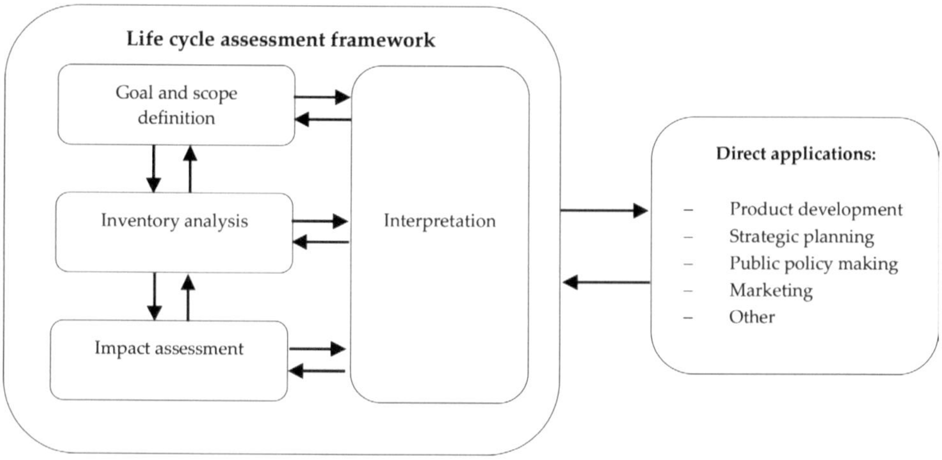

Figure 1.
Components of a life cycle assessment (LCA) according to International Organization for Standardization (Source: [9]).

Life cycle assessment (LCA) is one of the environmental management techniques, which aims to assess potential hazards to the environment of products, processes or entire systems. LCA as a tool to analyze waste management systems appeared in the early 1990s. It is worth noting that among researchers and decision makers, the use of LCA to analyze and develop waste management strategies has increased considerably over the last few years [2]. Moreover, LCA is a useful framework for assessing environmental performances [3]. The role of LCA has been increasing as it was proposed in many EU and Polish official documents [4]. Currently the LCA methodology is more and more frequently used as a tool for evaluating the environmental performance of products or services [5].

The LCA description is based on the ISO standard series 14040-14044 (2006) [6] and the guidelines provided by Guinée [7]. According to ISO, LCA is used for hot spot analysis, product or process improvement, comparative assertion, marketing and environmental policy.

In accordance with the ISO 14040 (2006) [7] standard, describing the principles and framework, LCA consists of the four phases [8] as illustrated in **Figure 1** [9]. Life cycle inventory (LCI), the second valuable step of LCA, is the most effective quantitative environmental assessment tool [10].

2. Methods

2.1 Goal and scope of the study

The goal definition describes the purpose of the study and the decision process to which it provides environmental decision support [8], and the scope includes the way the object of investigation is modeled. The functional unit and system boundaries are also determined at this step. The scope definition of an LCA study must address the following issues:

- the object of the study-functional unit;
- the system boundaries;
- the assessment criteria to be applied;

- the time scale of the study; and
- the technologies representing the different processes as presented by Hauschild and Barlaz [8].

Detailed issues (e.g., *functional unit*, *system boundaries*, *time scale of the study*, *technologies representing the different processes*) in scope definition of an LCA study is discussed by Hauschild and Barlaz [8].

2.2 Functional unit

The functional unit (FU), central concept in LCA [11, 12], is the measure of the performance delivered by the system under study [12], and definition of a FU is essential in LCA [13]. According to [8] for the LCA of waste management systems, the FU of the study could include:

- quantity of waste to be managed;
- composition of the waste;
- duration of the waste management systems; and
- quantity of the waste management (legal emission limits, requirement for residual products).

For the purpose of this study, a suggested FU is defined as amount of waste to be stored during the year—waste other than hazardous—and recovered and stored during the year at Kosodrza landfill (see **Figure 2**). Time coverage is year 2015.

2.3 Data quality

The problem of data quality in building an LCI, which is the foundation of any LCA [14], is discussed in [11]. Collection of LCI data is one of the most important

Figure 2.
LCI system boundary of the gate-to-gate for the MSW landfill considered in this study (source: photo from waste landfill in Kosodrza management communication).

stages in an LCA study [15]. Moreover, data quality is multidimensional and not necessarily quantitative [14]. LCI required a lot of data [12, 16] that are well correlated to the study context [14].

The paucity (reliability) of data can be a strong impediment in the conduct of LCA and explain the bias in choice of waste types to study [17].

The data used in this study involving consumption of energy and fuels, water, materials and waste are obtained from (i) site-specific measured or calculated data and (ii) secondary data taken from integrated permit issued by Marshal of the Podkarpackie region in Rzeszów for the establishment of municipal services in Ostrów (e.g., **Figure 3**) by entering the records concerning the waste landfill in Kosodrza (e.g., **Figure 4**), dated October 31, 2015, *and its subsequent amendments* [18]. Integrated permit has been issued *at the request of the interested party*.

The present LCI, as mentioned above, is representative for year 2015 by application of PN-EN ISO 14040:2009 [19].

A full publication of the inventory data used in this study is documented in [18]. In this case study, the system evaluated does not include anything upstream from the waste landfill operation.

As this study was based on the deterministic approach to LCI, uncertainty analysis was not carried out. However, very few assessments include effects of the waste composition, and waste LCAs often rely on poorly justified data from secondary sources, and uncertainty on LCA results associated with selection of waste composition data have been performed [20]. The LCI model can be used in full LCA study.

Figure 3.
Waste landfill in Kosodrza, in the community of Ostrów (source: [18]).

Figure 4.
Landfill for waste other than hazardous and inert wastes with separate hazardous waste facilities containing asbestos in Kosodrza (source: BIP based on https://www.google.pl/maps).

3. LCI of a modern MSW landfill

According to [20] the composition of waste materials has fundamental influence on environmental emissions associated with waste treatment, recycling and disposal and may play an important role also for the LCA of waste management solutions.

According to [21] to carry out a LCA, there is a need for LCI data in order to ensure a representative assessment. Major LCA methodological steps, including among others inventory analysis, are illustrated in [1], based on EC [22, 23]. LCI data on waste management processes involves recycling, source separation, collection, transport and upgrading of recyclables, and it is readily available [24].

Several definitions of solid waste exist. In the review given in [3], waste, according to [25], is neither water (wastewater) nor airborne (flue gases). According to [26] urban solid waste is defined as the waste generated by household, businesses, industries, institutions and markers, as well as the waste coming from the cleaning of streets and public areas [26].

It should be noted that Environmental Research and Education Foundation (EREF), a non-profit organization, is one of the largest sources of funding solid waste research in North America; it defines solid waste as [27]:

- municipal solid waste (e.g., residential, commercial, institutional);
- construction and demolition debris;
- certain industrial wastes (e.g., exploration and production waste, coal ash); and
- other wastes typically managed by the solid waste industry or generated by the public not included in the above list (e.g., electronic waste, disaster debris, etc.).

Agricultural wastes (that are not handled by the waste industry), nuclear waste and land-applied wastewater treatment sludge are generally not included in this definition [27].

4. LCI of municipal solid waste (MSW) management systems in Kosodrza, community of Ostrów

The consumption of energy and fuels, water, chemicals and waste obtained from secondary data taken from integrated permit issued by Marshal of the Podkarpackie region in Rzeszów for the establishment of municipal services in Ostrów by entering the records concerning the waste landfill in Kosodrza is given in **Tables 1–4**, respectively.

The maximum amount of waste to be disposed of through storage during the year will be:

- hazardous waste: 3000 Mg/year (10 Mg/day);
- nonhazardous waste: 156,393 Mg/year (500 Mg/day); and
- in the event of a situation deviating from the normal one—an additional 18,000 Mg/year of other waste.

Total amount of waste accepted for processing in landfill recovery processes is:

- the total amount of waste recovered in R5 processes per year will amount to 20,030 Mg/year (the amount of waste used to build inert layers on the landfill cannot exceed 6700 Mg/year); and
- the total amount of waste subjected to recovery in the R3 process will amount to 12,450 Mg/year during the year.

4.1 The leachate process

The integrated permit was issued for the operation of installations for the disposal of nonhazardous and inert waste with the capacity to receive more

No	Specification		Unit	Amount value
1	Gas oil		Mg/year	117
2	Tap water	Technological utilization	m^3/year	1000
		Sanitary utilization	m^3/year	75
3	*Chemical reagents* used for the reverse osmosis process purification plant	disinfectant sanitizer	Mg/year	1.5
		Sulfuric acid	Mg/year	100
		Hydrated lime	Mg/year	40
		Chlorinated lime	Mg/year	0.2
		Citric acid	Mg/year	3.0
		Sodium hydroxide	Mg/year	20
4	Electric power		kWh/ year	300,000
5	Hard coal		Mg/year	3.7

Table 1.
Type of energy, water, chemicals, and fuels—landfill for waste other than hazardous and inert wastes with separate hazardous waste facilities containing asbestos in Kosodrza.

No	Type of waste	Quantity of waste
1	Inorganic wastes	1000
2	Furnace linings and refractories from non-metallurgical processes	23
3	Other wastes	70
4	Mixed wastes from construction, renovation and dismantling	6000
5	Non-composted municipal solid waste	20,000
6	Other unused waste (waste from the mechanical and biological treatment plant)	50,000
7	Digested wastes of anaerobic decomposition of municipal solid waste	2003
8	Screenings	3000
9	Content of sand traps	2000
10	Sludges from non-biological treatment of industrial wastewater	4000
11	Solid wastes from preliminary filtration and screenings	500
12	Glass	2000
13	Other wastes (including mixed substances and objects) from mechanical treatment of waste	50,000
14	Other non-biodegradable waste	6000
15	Waste from marketplaces	3000
16	Sludges from septic tanks used to collect impurities	800
17	Waste from sewer manholes	800
18	Municipal waste not included in other subgroups	7000
19	Insulation materials containing asbestos	3000
20	Construction materials containing asbestos	3000

Table 2.
Types and amount of waste to be stored during the year—waste other than hazardous (all values in Mg/year).

No	Type of waste	Quantity of waste
1	Waste sands and loams	100
2	Waste resulting from cutting and rock cutting	100
3	Slag, bottom ash and boiler dust	300
4	Fly ash from coal	100
5	Defective ceramics, bricks, tiles and building ceramics (after thermal processing)	200
6	Worn (used) tires	200
7	Waste of concrete and debris from demolition and renovation	1000
8	Brick rubble	1000
9	Wastes of other ceramic materials and equipment items	560
10	Mixed or segregated waste from concrete, brick rubble and waste ceramic materials	3000
11	Plasters removed	500
12	Concrete parts and aggregates not containing asphalt	100
13	Soil and soil, including stones	5000
14	Dredging spoil	80
15	Torn rubble (aggregate)	40

No	Type of waste	Quantity of waste
16	Construction materials containing gypsum	200
17	Compost not meeting the requirements (unsuitable)	10,000
18	Stabilized municipal sewage sludge	2200
19	Sludges from water clarification	100
20	Minerals (e.g., sand, stones)	2500
21	Other non-specified fractions collected selectively (ashes and slags)	200
22	Soil and soil, including stones	2000
23	Waste from cleaning streets and squares	3000

Table 3.
Type and quantity of waste recovered during the year (Installation - Landfill for waste other than hazardous and inert wastes with separate hazardous waste facilities containing asbestos in Kozodrza - all values in Mg/year).

No	Type of waste	Quantity of waste
1	Packaging made of paper and cardboard	440
2	Plastic packaging	1200
3	Wood packaging	600
4	Metal packaging	440
5	Multi-material packaging	230
6	Glass packaging	1200
7	Packaging from textiles	220
8	Paper and cardboard	1500
9	Ferrous metals	600
10	Non-ferrous metals	500
11	Plastics and rubber	4000
12	Glass	2000
13	Other wood	500
14	Textiles	40
15	Other wastes (including mixed substances and articles) from mechanical treatment of waste containing dangerous substances	110
16	Other wastes (including mixed substances and articles) for mechanical processing of waste—oversize fraction with a grain size greater than 80.0 mm with the properties of combustible waste—preRDF	21,000
17	Other wastes (oversize fraction with a grain size greater than 80.0 mm—ballast)	14,000
18	Other wastes (biodegradable fraction)	25,000
19	Compost not meeting the requirements (not suitable to be used)	11,250
20	Other unmentioned waste (sieve fraction from stabilizer screening)	13,750
21	Other wastes (including fiberboard, leftover wood contaminated plastic)	700
22	Waste of concrete and debris from demolition and renovation	20,000
23	Brick rubble	2000
24	Iron and steel	100
25	Worn (used) tires	220

No	Type of waste	Quantity of waste
26	Other engine, gear and lubricating oils	1.8
27	Sorbents, filter materials, wiping cloths (e.g., rags, dishcloths) and protective clothing	0.3
28	Oil filters	0.2
29	Other unlisted items (air filters)	0.2
30	Worn out (used) devices containing hazardous elements	0.3
31	Lead-acid batteries and accumulators	0.2

Note: The waste mentioned in No. from 1 to 18 will be generated as a result of processing in the installation for mechanical waste treatment, the waste mentioned in No. 19 will be generated as a result of processing in the installation for biological waste treatment, the waste mentioned in No. from 20 to 23 will be generated as a result of the operation of the large-size waste disassembly point, the waste mentioned in No. from 22 to 24 will be generated as a result of crushing construction debris, the waste mentioned in No. 25 will be generated as a result of the plant's ongoing operation (arising as part of its current operation, machinery and equipment) and the waste mentioned in No. from 26 to 31 will be generated in connection with maintaining the efficiency of installations for mechanical and biological waste treatment.

Table 4.
Types and quantities of waste to be generated during the year and the source of waste generation (all values in Mg/year).

than 10 tonnes of waste per day and a total capacity of over 25,000 tonnes, with separate asbestos-containing hazardous waste units in Kosodrza, Ostrow commune (see **Figure 4**).

Description of the current installation and method of purification/pretreatment of the leachate in the landfill from the integrated permit is given below. The leachate process is performed in the two leachate tanks:

- leachate tank named ZRO1; and
- leachate tank named ZRO2.

The ZRO1 leachate tank is used to retain leachate arising within the existing quarters No. 1–8; it can be used to pump out leachates from the ZRO2 reservoir, i.e., from quarters No. 9, 10, 11, and 12.

Earth tank, insulated with 1.0 m thick, surface reinforced with a wreath and reinforced concrete grate, filled with openwork plates. The walls of the tank were made of grids made of reinforced concrete beams, 30 × 30 cm, creating grid structures over the bottom. Grill grates and slopes above the crown were secured with openwork concrete tiles 100 × 75 × 12.5 cm, on a geotextile with a weight of 400 g/m^2 and densified ballast made of gravel material. The bottom of the tank is a 20-cm-thick reinforced concrete slab.

ZRO2 reservoir located in the north-western part of the land designated for the extension of the landfill in the resulting triangle between the existing quarters No. 8, the factory road to quarters No. 9–12 (e.g., **Figure 5**) and A1 and A2 and the area of the leachate treatment plant.

Terrain open tank protected escarpments and the bottom triangular in plan. The structure of the tank bottom and walls will be sealed with a 1.0 m thick layer, 1.5 mm thick foil and geotextile g = 400 g/m^2, reinforced with a concrete construction.

The ZRO2 tank is the main retention reservoir for leachate from quarters No. 9–12. The leachate from quarters No. 9–12 will flow gravitationally to the P6 pumping station, from where they will be pumped into the ZRO2 leachate retention reservoir.

Figure 5.
New quarter No. 12 (source: [18]).

4.2 Container sewage wastewater treatment plant (CSWTP)—leachate treatment plant in reverse osmosis technology

Container sewage wastewater treatment plant (CSWTP) with a capacity of 30 m^3/d includes system called a single-stage membrane process ensuring obtaining the leachate parameters enabling them to be safely transported to the municipal sewage treatment plant. The treatment plant operates on the basis of the reverse osmosis process (e.g., **Figure 6**), the essence of which consists in passing the leachate from the storage site through a semipermeable membrane under the influence of the pressure set on the inlet side of the effluent.

Membrane separation is a purely physical separation; separated components do not undergo any chemical or biological transformation. The applied solution is a pilot solution for cooperation with the existing pretreatment plant in order to increase the cleaning effects. The effect of the treatment plant is to obtain purified leachate (permeate) and leachate residue (concentrate).

The sewage treatment plant works in a continuous system and cleanses the effluents from the ZRO2 reservoir, i.e., from quarters 9 to 12. The treatment plant will be controlled by means of a computer program and will work in an automatic system. The computer will monitor, through systematic conductivity measurement, the quality of treated leachate discharged into the environment. If the conductivity

Figure 6.
The treatment plant based on the reverse osmosis process (source: [18]).

rises above the programmed value, the installation will automatically stop, and a cleaning program for filters or modules will start.

Sewage plant is located in a paved square for turning vehicles. Container of treatment plant with dimensions of 12.2 × 2.5 m is set on a separate foundation.

The container is made of steel construction with a layer casing made of trapezoid sheet metal from the outside and a polypropylene plate from the inside of the container. The addition between the layers is a mineral wool insulation layer. Tight floor is made of chemically resistant material. The container has mechanical ventilation.

5. Literature review

The LCA literature on waste treatment can be found in [3]. According to [28], the annual total solid waste generation worldwide is approximately 17 billion tonnes, and it is expected to reach 27 billion by 2050 [1, 17]. Based on [29] in this amount, about 1.3 billion tonnes are currently municipal solid waste generated by world cities, which are anticipated to generate up to 2.2 billion tonnes by 2025 primarily due to population growth, increasing urbanization and socio-economic development of low- and middle-income countries [1]. The waste management problem in the EU is characterized by increasing per capita production of waste materials, the need for high levels of investment in physical infrastructure (incinerators, recycling facilities and landfills), institutional barriers, a wide range of stakeholders and a dynamic policy arena.

In this section we describe several studies with numerous examples demonstrating the waste management. Ref. [3] illustrates development of the regionalised municipal solid waste incineration model in France, which can be adapted to regional characteristics and incineration conditions in order to provide the best representation and most accurate predictions of MSW incineration in a given geographic area [3]. The world's largest center for urban waste by 2007, according to [30], was operational in Amsterdam in the Netherlands. This includes the city's sewage treatment plant and the expanded waste-to-energy plant for solid waste (SW) [30].

Details about Latin America, as a region strongly affected by the lack of equality in income distribution and big differences in the quantity of the waste generated daily and in its composition, can be found in [26].

Moreover, according to work presented by Savino [26], the regional assessment report on municipal solid waste management (MSWM), published by Pan American Health Organization (PAHO) in 2005, says: "The composition of waste in Latin America, although it varies among the different centers of population, maintains a strong component of foodstuff waste, with average values from 50 to 70% in weight, while around 25% of waste components is made up of paper, cardboard, metal, textile, leather, rubber and wood." According to studies carried out by national member International Solid Waste Association (ISWA) in Argentina, presented in [26], the percentages are as follows:

- adequate final disposition of SW in the metropolitan area of Buenos Aires at sanitary landfill 45%;

- the rest of Argentina 55%;

- adequate final disposition in sanitary landfill 10%;

- waste disposal in controlled sites 10%; and

- uncontrolled open-air dumps 35% [26].

The composition of waste in Buenos Aires is presented and shown in **Figure 7**.

The case of the sanitary landfill in Buenos Aires is illustrated in **Figure 8**.

In paper [31] a systemic approach for MSWM at both the household and the non-household level has been developed. It summarizes state-of-the-art available tools and compiles a set of guidelines for developing waste management master plans at the municipal level, and it provides a framework in the MSWM field for municipalities in Greece and other countries facing similar problems under often comparable socio-economic settings [31] [ZOTOS]. Moreover, the Hellenic State has defined sufficiently the legislative and political framework for MWSM, in frame of related EU legislative approaches, and the 4R (reduce-reuse-recycle-recover) concept is well promoted by the "National Planning of SWM"(Hellenic) constituted of two Joint Ministerial Decision, legislated in 1997 and 2000, respectively [30]. It is interesting to note that SWOT analysis is performed for MSWM (e.g., [31]).

In China landfill density cannot be as high as in developed countries because its population distribution and economic development are quite different [32]. The amount of MSW collected by local authorities in China has increased in parallel with rapid urbanization. The average rate of increase in the amount of MWS collected annually is about 6% [32]. Moreover, the overall status of

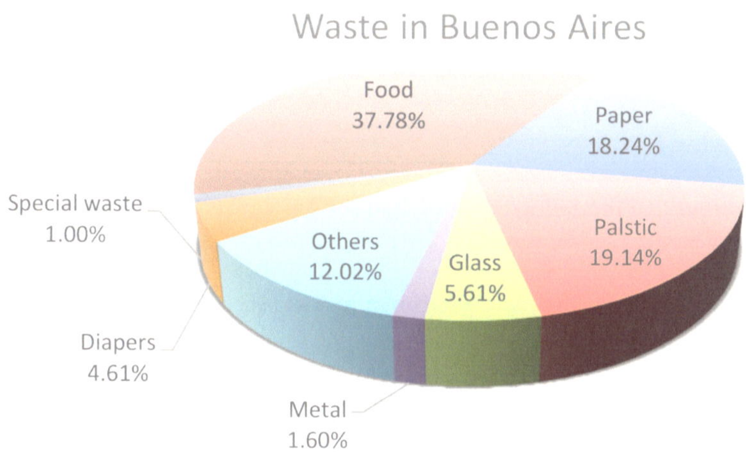

Figure 7.
Composition of waste in Buenos Aires (source: [26]).

Figure 8.
Sanitary landfill Norte III in Buenos Aires (source: [26]).

Figure 9.
Landfill site in Beijing City (source: [32]).

MSW treatment in China is still at the developing stages, with waste collection going from incomplete to complete collection and waste treatment going from decentralized disposal to sanitary landfilling [32]. Landfill site in Beijing City is presented in **Figure 9**.

6. Conclusions

The present LCI modeling of municipal solid waste management (MSWM) systems in Kosodrza, community of Ostrów, Poland case study was given according to PN-EN ISO 14040.

This study is focused on the operational results recorded in 2015, as defined in the goal and scope.

It should be noted that LCI work was performed using the secondary data obtained from integrated permit legislated for waste landfill in Kosodrza, community of Ostrów in Poland.

The results may be useful for MSWM in Poland. In the methodological approach regarding databases, boundaries were transparent and fully documented. Moreover, the results of this study can help MSW management authorities and practitioners to solve environmental and technical aspects and decision makers to understand the nature of the LCA. In addition to LCI, these data can be used to assess life cycle impact assessment (LCIA) as the next step of LCA methodology, and finally, full LCA should be conducted. The LCIA provides the analysis of collected data to evaluate contributions to various environmental impact categories. The final LCA of a modern MSW landfill should include the uncertainty of waste compositions.

The LCI study allows to identify and understand LCA approach from the view of further research work with a view to reduce the negative impact of *waste on* the environment as well as to reduce the negative impacts on ecosystems, on human health or on natural resources.

However this study has examined a case at the country level. This case study could be used by other domestic and international LCA studies of solid waste management systems.

The results obtained from this study can move the LCI on the waste management process one step forward and will assist in developing environmental awareness in the development of the National Waste Plan.

Acknowledgements

The authors wish to thank Marek Gubernat (Municipal Services Office in Ostrów) and Henryk Stolarczyk (Municipal Services Office in Ostrów) for supplying data on the MSWM in Ostrów. Relevant figures are reprinted with given sources. This publication and research was funded from subvention for the maintenance and development of research potential granted to AGH University of Science and Technology, Management Department, Kraków, Poland.

Conflict of interest

The authors declare that they have no conflict of interest.
The research does not involve human participants and/or animals.

Author details

Dariusz Sala* and Bogusław Bieda
Management Department, AGH University of Science and Technology, Kraków, Poland

*Address all correspondence to: dsala@zarz.agh.edu.pl

References

[1] Laurent A, Clavreul J, Bernstad A, Bakas I, Niero M, Gentil E, et al. Review of LCA studies of solid waste management systems—Part II: Methodological guidance for a better practice. Waste Management. 2014;**34**:589-606

[2] Björklund A, Finnveden G, Roth L. Application of LCA in waste management. In: Christensen TH, editor. Solid Waste Technology & Management. Vol. 1. Chichester: Blackwell Publishing Ltd.; 2011. pp. 137-160

[3] Déchaux C, Nitschelm L, Giard L, Bioteau T, Sessiecq P, Aissani L. Development of the regionalised municipal solid waste incineration (RMSWI) model and its application to France. International Journal of Life Cycle Assessment. 2017;**22**:1514-1542

[4] Nunez P, Jones S. Cradle to gate: Life cycle impact of primary aluminium production. International Journal of Life Cycle Assessment. 2016;**21**:1594-1604

[5] Grzesik K, Bieda B, Kozakiewicz R, Kossakowska K. Goal and scope and its evolution for life cycle assessment of rare earth elements recovery from secondary sources. 17th International Multidisciplinary Scientific GeoConference SGEM 2017, www.sgem.org, SGEM2017 Conference Proceedings, 29 June–5 July, 2017. Vol. 17(41). pp. 107-114. DOI: 10.5593/sgem2017/41/S18.014

[6] International Organization for Standardization (ISO). Environmental management—Life cycle assessment—Principles and framework. ISO 14040:2006. 2006. Available from: https://infostore.saiglobal.com/en-gb/Standards/ISO-14040-2006-R2016--614392_SAIG_ISO_ISO_1409806/. [Accessed 2019-01-28]

[7] Guinée JB, Gorrée M, Heijungs R, Huppes G, Kleijn R, de Koning A, et al. Life cycle assessment; An operational guide to the ISO standards; Part 3: Scientific Background. Den Haag and Leiden, The Netherlands: Ministry of Housing, Spatial Planning and Environment (VROM) and Centre of Environmental Science (CML); 2001. Available from: http://www.leidenuniv.nl/cml/ssp/projects/lca2/lca2.html [Accessed 2018-11-15]

[8] Hauschild MZ, Barlaz MA. LCA in waste management: Introduction to principle and method. In: Christensen TH, editor. Solid Waste Technology & Management. Vol. 1, 2011. Chichester: Blackwell Publishing Ltd; 2011. pp. 113-136

[9] Jungbluth N, Keller R, Doublet G, König A, Eggenberger S. Report on life cycle assessment, economic assessment, potential employment effects and exergy-based analysis: Part I—LCA. Deliverable 7.3. SUSMILK—Re-design of the dairy industry for sustainable milk processing, Seventh Framework Programme: Project no. 613589. Funded by EC. 2016. 2002;**20**:16-22

[10] Millet D, Bistagnino L, Lanzavecchia C, Camous R, Poldma T. Does the potential of the use of LCA match the design team needs? Journal of Cleaner Production. 2007;**15**:335-346

[11] Bieda B, Skalna I, Gaweł B, Grzesik K, Henclik A, Sala D. Life cycle inventory processes of the integrated steel plant (ISP) in Krakow, Poland—Continuous casting of steel (CCS) a case study. International Journal of Life Cycle Assessment. 2018;**23**:1274-1285

[12] Sonnemann G, Castells F, Schumacher M. Integrated Life-Cycle and Risk Assessment for Industrial Processes. CRC Press Company. Boca Raton, Fl, USA: Lewis Publishers; 2004. p. 392

[13] Gómez P, Elduque D, Sarasa J, Pina C, Javierre C. Influence of

composition on the environmental impact of a cast aluminum alloy. Materials. 2016;**9**(6):412. DOI: 10.3390/ma9060412

[14] Seto KE, Panesar DK, Churchill CJ. Criteria for the evaluation on life cycle assessment software packages and life cycle inventory data with application to concrete. International Journal of Life Cycle Assessment. 2017;**22**:694-706

[15] Finnveden G, Hauschild MZ, Ekvall T, Guinée J, Heijungs R, Hellweg S, et al. Recent development in life cycle assessment. Journal of Environmental Management. 2009;**91**:1-21

[16] Morales M, Quintero J, Aroca G. Environmental assessment of the production and addition of bioethanol produced from *Eucalyptus globulus* to gasoline in Chile. International Journal of Life Cycle Assessment. 2017;**22**:525-536

[17] Laurent A, Bakas I, Clavreul J, Bernstad A, Niero M, Gentil E, et al. Review of LCA studies of solid waste management systems—Part I: Lessons learned and perspectives. Waste Management. 2014;**34**:573-588

[18] BIP (Biuletyn Informacji Publicznej). Pozwolenie zintegrowane. 2014. Available from: http://www.bip.podkarpackie.pl/index.php/informacja-o-srodowisku/pozwolenia-zintegrowane/wydane/1522-gminny-zaklad-uslug-komunalnych-sp-z-o-o-w-ostrowie-skladowisko-odpadow-komunalnych-z-wydzielonymi-kwaterami-na-odpady-niebezpieczne-zawierajace-azbest-w-kozodrzy [Accessed 2018-12-8]

[19] PN-EN ISO 14040. Environmental Management—Life Cycle Assessment—Principles and Framework. 2009. Available from: http://sklep.pkn.pl/pn-en-iso-14040-2009p.html [Accessed 2018-12-8]

[20] Bisinella V, Götze R, Conradsen K, Damgaard A, Christensen TH, Astrup TF. Importance of waste composition for Life Cycle Assessment of waste management solutions. Journal of Cleaner Production. 2017;**164**:1180-1191

[21] Brogaard LK, Damgaard A, Jensen MB, Barlaz M, Christensen TH. Evaluation of life cycle inventory data for recycling systems. Resources, Conservation and Recycling. 2014;**87**:30-45

[22] European Commission (EC)—Joint Research Centre—Institute for Environment and Sustainability. International Reference Life Cycle Data System (ILCD) Handbook—General guide for Life Cycle Assessment—Detailed guidance. 1st ed. EUR 24708 EN. Luxembourg, LU: Publications Office of the European Union; 2010

[23] European Commission (EC)—Joint Research Centre—Institute for Environment and Sustainability. Supporting Environmentally Sound Decisions for Waste Management—A technical guide to Life Cycle Thinking (LCT) and Life Cycle Assessment (LCA) for waste experts and LCA practitioners. EUR 24916 EN. Luxembourg, LU: Publications Office of the European Union; 2011

[24] Merrild H, Larsen AW, Christensen TH. Assessing recycling versus incineration of key materials in municipal waste: The importance of efficient energy recovery and transport distances. Waste Management. 2012;**32**(5):1009-1018

[25] Christensen TH. Solid Waste Technology & Management. Chichester, UK: Blackwell Publishing Ltd.; 2011

[26] Savino AA. Large population – more sanitation. The state of urban solid waste management in Latin America. Waste Management World. 2008. Available from: https://waste-management-world.com/search?q=Savi

no+AA.+Large+population+%E2%80%93+more+sanitation.+The+state+of+urban+solid+waste+management+in+Latin+America. [Accessed 2019-01-28]

[27] EREF. Environmental Research & Education Foundation. 2018. Available from: https://erefdn.org/research-grants-projects/how-to-apply-for-grant/ [Accessed 2018-11-15]

[28] Karak T, Bhagat RM, Bhattacharyya P. Municipal solid waste generation, composition, and management: The world scenario. Critical Reviews in Environmental Science and Technology. 2012;**42**:1509-1630

[29] Hoornweg D, Bhada-Tata P. What a Waste—A Global Review of Solid Waste Management. Washington DC, USA: World Bank; 2012. Available from: https://siteresources.worldbank.org/INTURBANDEVELOPMENT/Resources/336387-1334852610766/What_a_Waste2012_Final.pdf Accessed 2018-12-8

[30] McCarthy T, Mulder M, van Berlo M. A symbiotic solution. Integrating urban waste and sewage management. Waste Management World. 2007. Available from: https://waste-management-world.com/a/a-symbiotic-solution. [Accessed 2019-01-28]

[31] Zotos G, Karagiannidis A, Zampetoglou S, Malamakis A, Antonopoulos IS, Kontogianni S, et al. Developing a holistic strategy for integrated waste management within municipal planning: Challenges, policies, solutions and perspectives for Hellenic municipalities in the zero-waste, low-cost direction. Waste Management. 2009;**29**(5):1686-1692. DOI: 10.1016/j.wasman.2008.11.016

[32] Haiyun X. All the waste in China – the development of sanitary landfilling. Waste Management World. 2008. Available from: https://waste-management-world.com/a/all-the-waste-in-china-the-development-of-sanitary-landfilling. [Accessed 2019-01-28]

Chapter 7

Urban Management Model: Municipal Solid Waste for City Sustainability

Claudia E. Saldaña Durán and Sarah Messina

Abstract

The population growth arises the increase of municipal solid waste production in urban areas causing daily hundreds of tons of waste. Moreover, its composition characteristics comprise toxic and polluting elements that require infrastructure and enormous local resources for its treatment. The final disposition of this waste is an important issue; it is the key element to control the environmental contamination of soil and pollution of local water sources. Urban Management Model: municipal solid waste for city sustainability, it is based on the Government-Society-Academia alliance. Through a social and technological approach, this model holds the importance of knowledge transfer and its connection with key social actors. The study opens several future alternative solutions such as: biotechnology, technological development, marketing and trading materials to be reused and recycled, special studies for the final disposition destinations, and studies of companies' organization. Essential elements to provide a solution for the high production of waste problem in cities were conducted.

Keywords: stakeholders, recycling, Government-Society-Academia, selective separation

1. Introduction

In a more globalized and urban world as well as environmentally deteriorated, it is suggested that around 60–75% of global population will live in urban areas during the period of 2025–2050 [1]. This approach leads to many problems in the urban environment, such as population concentration, shortage of housing, scarcity of resources, reduction of biodiversity, air, soil and water pollution [2].

The final disposal of the waste is a serious issue, since it is the key element to control the environmental contamination of soil and pollution of local water sources. In past and even today many Mexican cities have disposed of their municipal waste in an inappropriate manner, using uncontrolled landfills to bury their garbage, causing a chain of environmental degradation. Solid waste management is defined as the discipline associated with the control of generation, storage, collection, transfer and transport, processing and disposal of solid waste in a way that harmonizes with the best principles of public health, economics, engineering, conservation, esthetics and other environmental and public considerations [3]. Within this scope, all administrative, financial, legal, planning and engineering functions are included.

IntechOpen

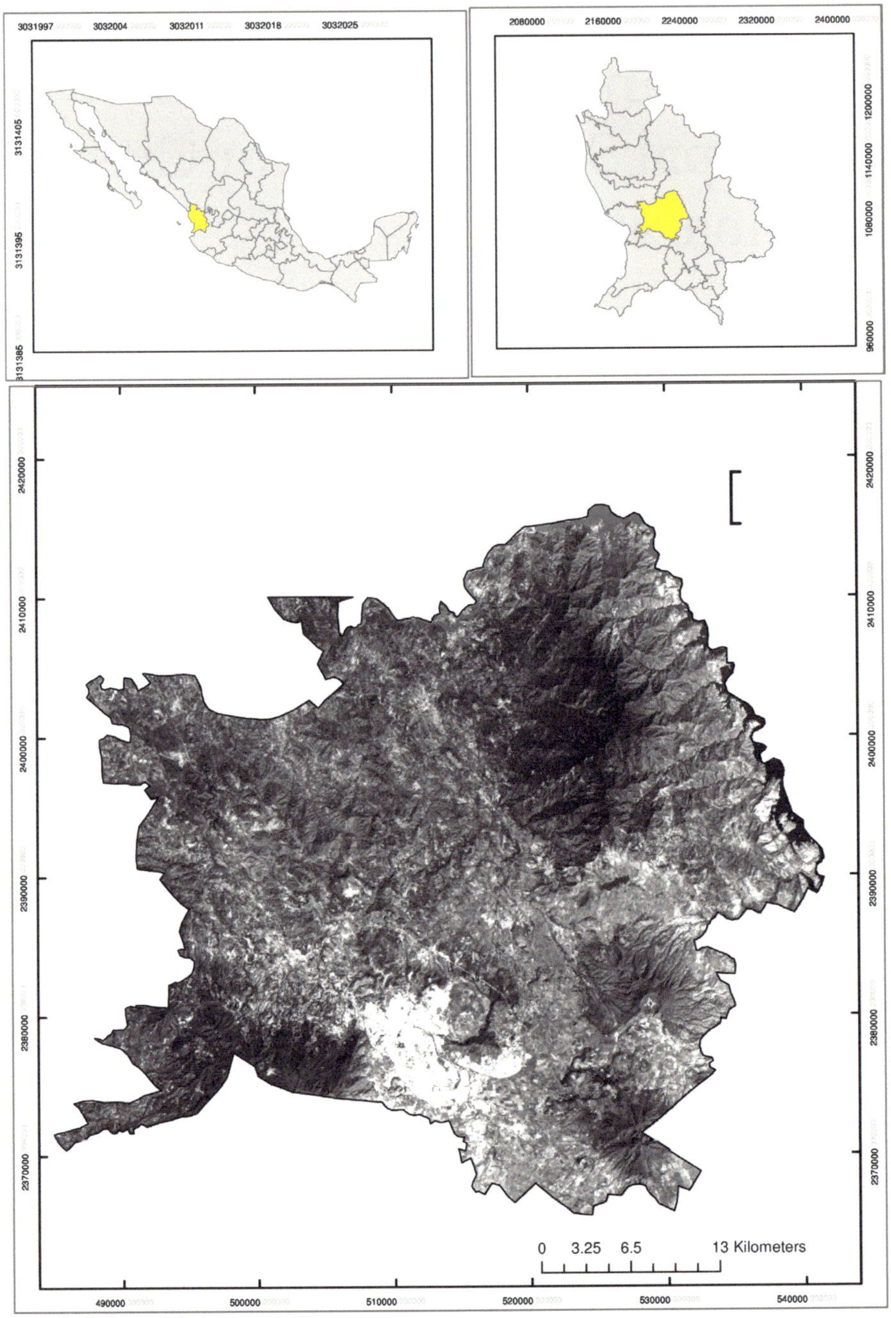

Figure 1.
Location of the study site, municipality of Tepic, Nayarit, Mexico.

In the last decades, Municipal Solid Waste (MSW) management systems have a complex and multifactorial behavior, due to large diversity of the materials that compose this waste, causing an environmental cost to cities. The technological, economic and environmental policies alternatives have caused changes in the waste management practice, which further complicates the scenario. Thus, it unleashes a new paradigm in the sustainable development of cities, at the local level but which impacts globally.

The sustainable management of Municipal Solid Waste requires a holistic approach that considers the parties involved, their relationships and different factors of complex decision making, in a sensible and logical way. In this chapter, we introduce an Urban Management Model: Municipal Solid Waste for city sustainability, which outline a multidisciplinary in middle size cities with a total population that goes from 100,000 to 1 million people and, a territory extension that goes from 1 to 5000 m^2, **Figure 1**.

In addition, it integrates the stakeholders in the individual and group decision making in the management of the MSW, as well as the social, economic, political and environmental aspects; for which it is necessary to establish their relationships and compare them.

Therefore, it is necessary to evaluate it through an ex-ante and ex-post multi-criteria analysis, regarding an environmental problem for decision making and to model it adequately, to draft well-structured strategies in the decision-making process for the scenarios future of the Municipal Solid Waste management system.

2. Contribution of the urban management model

Urban management model of urban solid waste is generated, applied and evaluated, based on a government-society-academy perspective. This model is designed through a social perspective and a technological perspective. The study considers four phases: key social actors, recycling, final disposal, public policies and model evaluation. This model postulates the incorporation of environmental management and sustainability through the link between local government-society and academia that will influence the plants and programs of solid waste (**Figure 2**).

2.1 Stakeholders

The main challenge in solid waste management is to develop models that help decision-making. Also considering cooperative interactions among stakeholders, which may be groups or individuals, that impact or are impacted by the USW

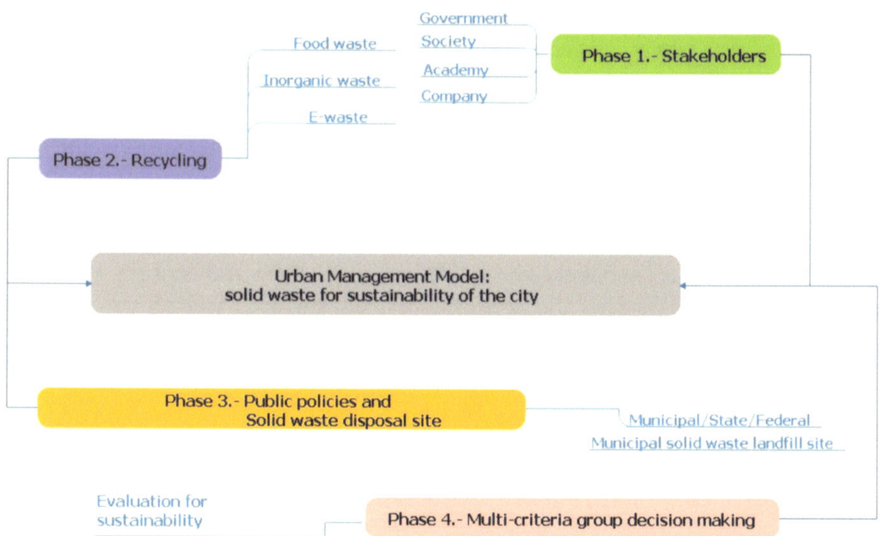

Figure 2.
Urban Management Model: municipal solid waste for city sustainability.

Stakeholder	Social actor	Function
Academia (A)	Academia and researchers	Management Knowledge transfer
Local Government (B)	Local Government of Tepic Department of Public Cleanliness Department of Ecology	Strengthen processes Operate plans and programs
Organized groups of Society (C)	Citizen Action Committees Environmental activists	Promote connection with the community Promote environmental culture among citizens

Table 1.
Stakeholders participating in the Urban solid waste management in the city of Tepic, Nayarit, Mexico.

system. It requires the participation of many experts and stakeholders for the waste reduction. It is therefore important to establish a link between all groups to produce a holistic approach to solid waste management [4–8].

In this study the stakeholders were convened to form a team: Government, Society and Academia. Following participatory and educational methodologies such as significant learning, different activities were designed to raise awareness, educate and train participants. In **Table 1**, participating stakeholders and its function are described.

2.2 Recycling of urban solid waste

The process of transformation of the materials obtained from the Municipal solid Waste was carried out in a first stage based on a mechanical treatment that is the grinding. In order to reduce volumes of compaction and market the products as raw materials. The recycling of organic solid waste was carried out through the study and quantification of food material to produce quality compost and its application in the region crops. The recycling of inorganic solid waste such as plastic, metal, paper and cardboard was made through the prototype development based on mechanical treatments. The recycling of electronic waste was carried out as a program to collect electronic equipment called "Recyclatron" in order to promote a culture towards sustainability in the management of electronic waste and to be a reference model for social and environmental responsibility within the community. The operation of the program was based on the logistics of the integrated management of urban solid waste, under the Official Mexican Standard NOM-161-SEMARNAT-2011 [9]. It involves five stages: (i) collection; (ii) characterization; (iii) quantification; (iv) recovery and reuse; (v) marketing and trading. Four editions were held, every 2 years, the program incorporates engineering students from the Autonomous University of Nayarit who supervise the operation and execution of the process, **Figure 3**.

2.3 Final disposal of urban solid waste

The growth of the population in urban areas and the development of cities should support environmental sustainability in Municipal Solid Waste management systems. The final disposal of this waste is a serious issue, since it is the key element for the control of environmental local soils contamination and water sources pollution. Therefore, the need to designate suitable sites for their final disposal is justified. Potential zones were identified for the location of a solid waste landfill in

Figure 3.
Stakeholders municipal solid waste for city sustainability.

the municipality of Tepic. A spatial analysis of the municipality was carried out, contrasting four criteria indicated in the Official Mexican Standard NOM-083-SEMARNAT-2003 [10], and two other natural features, slope and coverage, as well as land use. It was possible to identify a 5.4% surface of the municipality of Tepic (about 9090.8 ha) with appropriate land features for the location of a sanitary landfill; the remaining 94.6% has at least one characteristic that limits or restricts it for this purpose, **Figure 4**.

2.4 Public policies in the solid waste selective separation

Propose public policies with regulations towards actions of selective separation, incorporation of the informal cleaning sector to the formal one, through the system of consultation, studies and discussions with the community. The proposal considered a regulatory framework for actions to separate and treat solid waste in cities. The environmental legislation of different order was analyzed in the three levels of competence: federal, state and municipal. The following guidelines are proposed for its execution. Chapter one—General provisions—Establish the public interest and the regulations of the subject.—Establish what is intended to regulate—Conceptual framework of terms—Municipality competence—Cleaning service—Environmental education. Chapter two—Management—Authorities—Powers of the authorities-Inspection and surveillance. Chapter three—COMPREHENSIVE waste management—Solid waste and special management—Hazardous waste Chapter four—organization Service delivery—Sweeping system—Collection system—Transportation and transfer system—Treatment and final disposal. This comprehensive management policy

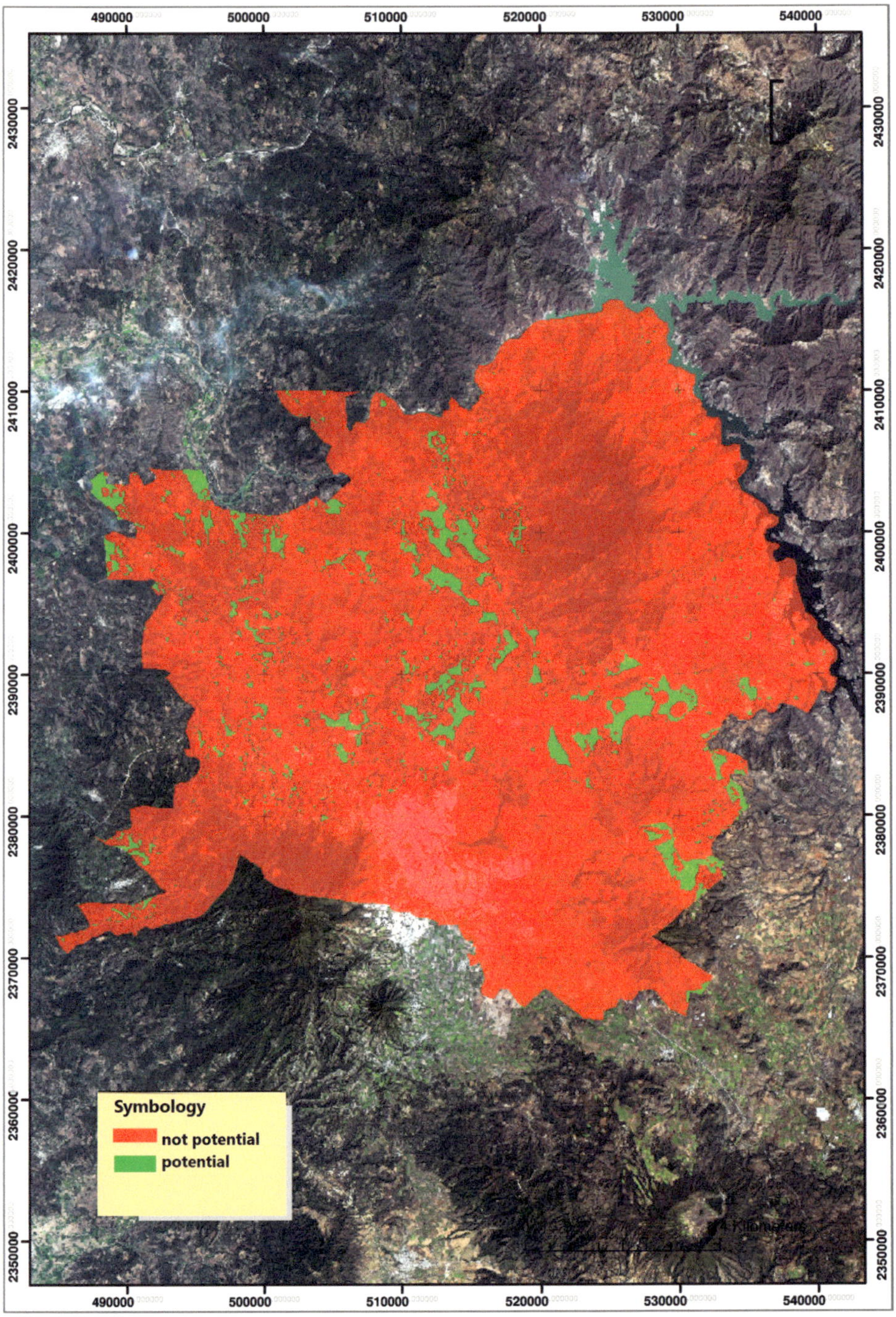

Figure 4.
Land potential for the location of landfill, municipality of Tepic, Nayarit, Mexico.

of urban solid waste presents the multilateral problem of waste management with a multi and interdisciplinary approach in order to solve it, including the legal, institutional, technical, economic, land-use planning and awareness, environmental education and participation of the citizenship. It is proposed for its fulfillment to present it before the State of Nayarit Congress for its study, revision or approval.

DOI: http://dx.doi.org/10.5772/intechopen.82839

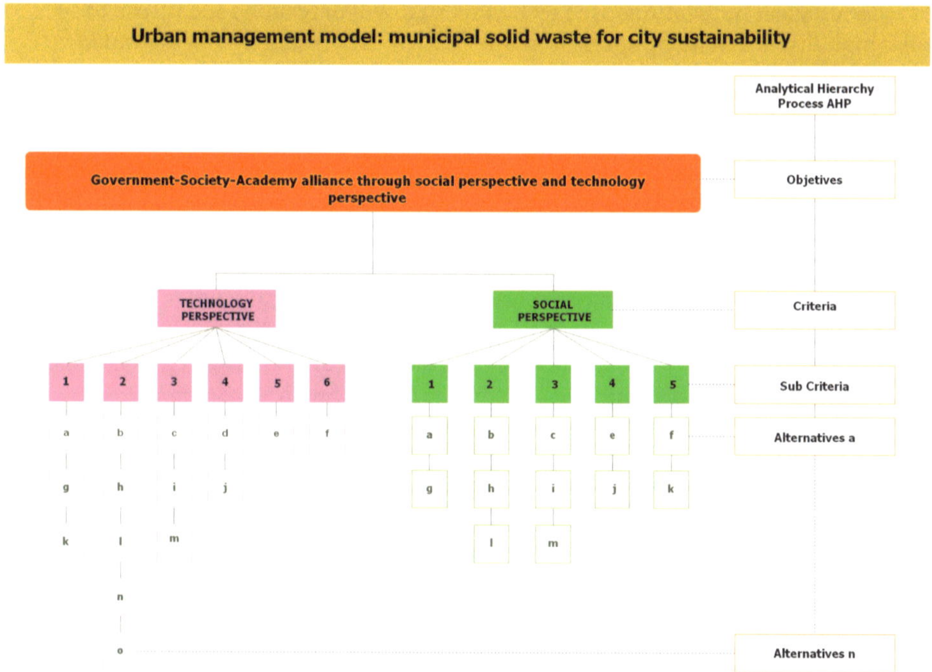

Figure 5.
Evaluation urban management model: municipal solid waste for city sustainability.

2.5 Evaluation of the management model

The proposed model was based on the stakeholders: Government-Society and Academia. Through a social and technological perspective. Taking the city of Tepic as study area. The methodology contribution used for its evaluation was the multi-criteria and/or multi- objective analysis, based on a set of techniques used in the multidimensional decision making to assess a group of alternatives which covers and satisfy one or several objectives, in terms of multiple criteria. These studies facilitate the balanced analysis of all stages of PLANNING problems, because several intangible effects, such as social effects and environmental repercussions can be fully considered [11]. In this study it was processed through the Analytical Hierarchy Process (AHP) to determine the contribution based on set of criteria in the project development. By the Analytical Hierarchy Process, the AHP, it is possible to organize the problem information, decompose it and analyze it by parts, visualize the variations presented when there are changes in each level of hierarchy and synthesize. The first step is to identify all the elements that intervene in the decision-making process and the levels at which these elements can be grouped in a hierarchical way. The criteria to assess: society, treatments, final disposition and public policies and generation and evaluation of the model, **Figure 5**.

3. Conclusions

The Urban Management Model: Municipal Urban solid waste for city sustainability establishes the indispensable elements to solve the problem of high production in cities. With a multidisciplinary vision, different disciplines are incorporated from engineering, social, marketing, local economic development, sustainability, and management of organizations. In this way, the development and dissemination of technological innovations are key factors that will determine the future of sustainability in a highly populated planet with an environmentally degraded urbanized surface.

The model allowed to achieve the transfer of knowledge, in all intervention spaces. At the levels of stockholding and in the technical processes: Academy-Government-Society and the incorporation of Companies, an adequate network of relationships were formed for the generation and transfer of knowledge. In addition, a strategic alliance was achieved. They were established the basis towards the study of biotechnology, the technological development, marketing of materials for recycling, space studies for final disposal sites, organizational studies in recycling companies incorporating the vision of environmental management systems. This range of possibilities raises the scope of the study and allows several paths in the investigation of applied science and the link with the business sector.

Acknowledgements

This research work has been supported by CONACYT through the project 247699 CONACYT-PDCPN-2014.

Conflict of interest

The authors declared no potential conflicts of interest with respect to the research, authorship, and/or publication of this chapter.

Author details

Claudia E. Saldaña Durán* and Sarah Messina
Universidad Autónoma de Nayarit, Tepic, Nayarit, México

*Address all correspondence to: cesduran@uan.edu.mx

References

[1] UN DESA United Nations, Department of Economic and Social Affairs, Population Department. World Urbanization Prospects. Vol. 12. United Nations; 2014. p. 32

[2] Khan D, Samadder SR. Municipal solid waste management using geographical information system aided methods: A mini review. Waste Management & Research. 2014;**32**(11):1049-1062. DOI: 10.1177/0734242X14554644

[3] Vergara SE, Tchobanoglous G. Municipal solid waste and the environment: A global perspective. Annual Review of Environment and Resources. 2012;**37**:277-309. DOI: 10.1146/annurev-environ-050511-122532

[4] Aleluia J, Ferrao P. Characterization of Urban waste management practices in developing Asian countries: A new analytical framework based on waste characteristics and Urban dimension. Waste Management. 2016;**58**:415-429. DOI: 10.1016/j.wasman.2016.05.008

[5] Cole C, Osmani M, Quddus M, Wheatley A. Towards a zero waste strategy for an English Local Authority. Resources, Conservation and Recycling. 2014;**89**:64-75. DOI: 10.1016/j.resconrec.2014.05.005

[6] Karmperis AC, Aravossis K, Tatsiopoulos IP, Sotirchos A. Decision support models for solid waste management: Review and game-theoretic approaches. Waste Management. 2013;**33**(5):1290-1301. DOI: 10.1016/j.wasman.2013.01.017

[7] Moh YC, Latifah AM. Solid waste management transformation and future challenges of source separation and recycling practice in Malaysia. Resources, Conservation and Recycling. 2017;**116**:1-14. DOI: 10.1016/j.resconrec.2016.09.012

[8] Yu H, Solvang WD. An improved multi-objective programming with augmented ε-constraint method for hazardous waste location-routing problems. International Journal of Environmental Research and Public Health. 2016;**13**(6):548. DOI: 10.3390/ijerph13060548

[9] NOM-161-SEMARNAT-2011 Official Mexican Standard. That Establishes the Criteria to Classify Waste of Special Management and Determine which are Subject to the Management Plan; the List of them, the Procedure for the Inclusion or Exclusion to Said List; as well as the Elements and Procedures for the Formulation of the Management Plans

[10] NOM-083-SEMARNAT-2003 Official Mexican Standard. Environmental Protection Specifications for the Selection of the Site, Design, Construction, Operation, Monitoring, Closure and Complementary Works of a Site for the Final Disposal of Urban Solid Waste and Closure and Complementary Works of a Site for the Final Disposal of Urban Solid Waste and Special Management

[11] Saaty TL. Decision Making. The Analytic Hierarchy Process. N.Y: McGraw-Hill Book Co; 1980. p. 287

Chapter 8

Decentralization and Solid Waste Management in Urbanizing Ghana: Moving beyond the Status Quo

Richard Kyere, Michael Addaney and Jonas Ayaribilla Akudugu

Abstract

Waste management is competing with more pressing economic and social issues such as social protection programs, education, and health. The government of Ghana has therefore decentralized the waste management system in the country. With this development, local government authorities and private sector actors are now playing key roles in waste management in the country. This study sought to examine decentralized solid waste management in the Berekum and Dormaa Municipalities in the Brong Ahafo Region of Ghana. Specifically, it analyzed the involvement of the private sector in solid waste management, and the quality of waste management services in the two selected municipalities. Through a survey of 312 households, the study analyzed the performance improvement, regulatory policy, and sustainable service delivery of solid waste management in the municipalities. The study found that there were no mechanisms for full cost recovery to include majority of the residents, who patronize communal collection service. The study therefore recommends the adherence to normative standards and agreed rules, adoption, and use of appropriate cost recovery strategies for low-income groups as well as the restructuring of institutional arrangements to ensure user involvement and enforcement of legislation to improve municipal solid waste management in Ghana.

Keywords: decentralization, municipal solid waste management, municipal authorities, private sector, urbanization

1. Introduction

Waste management remains a major challenge to management governments in Africa. In Ghana, the increasing rate at which waste is generated in the cities is alarming; yet government has not been able to respond in an equal measure. The proportion of populations living in urban areas in Africa is expected to increase from 40% in 2010 to about 57% in 2050 [1]. This incomparable rise in the level of urbanization in the first half of the twenty-first century Africa has goaded a variety of questions, apprehension, and agitation about the possible connotations of this development on the quality of life of Africa's rising population, and for environmental health in general [2]. This phenomenon has unquestionably buoyed a proclivity to consider the twenty-first century as marshaling in a period of predominantly urban civilization in Africa where urbanism is rapidly dominating ruralism [3]. Yet, the challenge of considerable transformation in the manner and pattern of urban functionality becomes ever more complex in the midst of unsustainable waste management problems [4].

The last three decades have seen a tremendous shift in government policies toward decentralization in the developing world. These policies are typically a component of comprehensive process of political, economic, social and technical reforms [5]. This has been inspired by new efforts of democratization and process of 'modernization' of the state. It can be argued that these initiatives combined to foster accountability, cost consciousness and competition in the public sector as well as develop a new role for the state in enabling and regulating rather than taking the place of the private sector. On the flipside, solid waste management (SWM) has become an important part of the urban environment as well as the planning of the urban infrastructure to safeguard a safe and healthy human environment. Continuous urbanization of developing countries at a very high rate has created serious problems of waste disposal as a result of uncontrolled and unmonitored urbanization [6]. Waste is a continually growing problem at the global, regional and local levels. The World Bank [7, 8], reported that there will be 70% increase in urban solid waste globally with a projected rise in the amount of waste, from 1.3 to 2.2 billion tonnes per year from 2012 to 2025, which will lead to a rise in the annual global costs of global waste from $205 billion to $375 billion. Within the same 13 years span, developing countries are facing the greatest challenges in the waste management sector.

In Africa, the poor state of solid waste management in urban areas is not only an environmental problem but also a major social handicap. In Kenya, it is expected that the amount of solid waste generated will increase from 2000 to 10,171 tonnes per day by 2025 [9]. The problem is further aggravated by the lack of financial as well as technical expertise in SWM technology and management especially in the sphere of collection, transportation, processing and final disposal. Whereas aspects like recycle, reuse and recovery of the solid waste is disorganized in most cases. In this context, the responsible persons or agencies concerned with public health and environment protection face the crisis of ineffective SWM. In the Ghanaian context, the situation is not different. Due to rapid urbanization, Ghana's major agglomerations have been growing quickly but have lacked a concurrent expansion in SWM. Addaney and Oppong [4] observe that Ghana like other developing countries has over the years had difficulties in municipal solid waste management with regards to infrastructural and technical inefficiencies. In view of this, the government has attempted to decentralize the SWM service delivery. These efforts have often become embroiled in politics, with less emphasis on efficient SWM delivery. In isolated cases where services have been decentralized, there have been inadequate policy direction and limited resource transfers to the lower levels of governance. Consequently, effective decentralized solid waste management has not been forthcoming.

The solid waste management subsector has been bedeviled with ineffectiveness despite the adoption of a number of policies and reform programs. Principally, decentralization has been designed to ensure efficiency and better service delivery at the local level. Despite this, there still exist challenges such as asymmetrical waste collection, waste overflow from bins, inadequate storage containers, and disposal of waste in unauthorized space in most municipalities in Ghana [4]. These challenges lead to public health hazards, esthetic nuisance, and environmental pollution. The public health implications have been fazing, accounting for about 5% of the GDP [10]. Data from the Ghana Health Service indicate that six (6) out of the top ten (10) diseases in Ghana are linked to poor environmental sanitation, with malaria, diarrhea and typhoid fever jointly constituting 70–85% of out-patient cases at health facilities [10]. The Berekum and Dormaa Municipalities are no exception to these undesirable environmental problems.

Therefore, this study attempts to unpack the difficulties face by the municipal authorities in keeping pace with solid waste facilities development and management. It is driven by the question of how decentralized SWM has evolved to ensure quality and sustainable service delivery in the medium-size towns of Berekum and Dormaa in the Brong Ahafo Region of Ghana. It analyzed the evolving practices of decentralized solid waste collection for sustainable service delivery, the service quality of decentralized urban SWM institutions; and the factors which explain the differences in service quality of the decentralized SWM of the two municipalities. It adopts the assertion that the inability of municipal authorities to effectively manage their solid waste usually leads to inefficient use of time and resources, and which eventually leads low productivity and poor service quality [4, 11].

2. Municipal solid waste management: theories and normative practices

Generating solid waste (SW) is inevitable. Cities in developing countries have frequently been unable to keep up with the provision of basic services [12]. About 40% of the solid waste generated in developing countries is uncollected, piles up on streets and in drains, contributing to flooding and the spread of disease. In addition, domestic and industrial effluents are often released into waterways with little or no treatment [12]. Solid waste has been a major challenge for municipal authorities for about 6000 years now [13]. The concept of waste is relative in two main respects. Firstly, something becomes waste when it loses its primary function for the user. Hence, one person's waste output is often someone else's raw material input. Secondly, the notion of waste is also relative to the technological state of the art and to the location of its generation ([14], p. 70). Waste is therefore a very dynamic concept and must be looked at within these two contexts. Many transnational organizations including the United Nations Environment Programme (UNEP) have their own definitions to the notion of waste. The UNEP [15] defined waste as any substances or objects which are disposed of or are intended to be disposed of or are required to be disposed of by national law.

Wastes that are solid are termed to as "refuse" or solid waste [16]. Waste has been defined differently by many Authors with different meaning. One definition is that waste is 'unwanted' by the first user. It is therefore anything that is no longer 'unwanted' dependent on the time and the prevailing circumstances. Solid waste today is increasingly defined as "natural resources out of place" or as "new materials for technologies not yet found" [17]. Many governments now regard waste as a useful source of income and as such policies have been geared toward this potential by both the government and the public sector to harnessing this potential. The recycling subsector for example, is an essential industry generating revenues and jobs for a larger number of people in the world today. Waste Watchers [18] defined solid waste management as everything that must be done to handle all the solid waste produced in a community, including collecting, transporting, processing and disposal of waste. Similar to this is the one put forward by Tchobanoglous et al. [19] that SWM involves the collection, treatment and disposal of non-hazardous waste.

Waste generation is the most important aspect to look at in order to have effective SWM system. The generation of waste varies considerably between countries based on the culture, public awareness and management [20, 21]. Waste generation comprise those activities in which materials are identified as no longer of any value by the owners/users and either thrown away or gathered for disposal [22]. Generally, developed countries generate more waste than developing countries [23]. Countries in Asian and African region produce waste in the range of 0.21–0.37 tonnes/capita/year, while European countries generate higher amount of waste with 0.38–0.64 tonnes/capita/year [24]. The waste generated by a population is a function of consumption patterns and thus of socioeconomic characteristics and the interest in and willingness to pay for collection services ([25], p. 35).

Disposal is broadly defined to include the collection, storage, treatment or processing, utilization, or final disposal of waste. It involves the process of getting rid of the waste materials that people generate [26]. Information on waste generation is important to determine the most suitable waste disposal options. The main purpose in implementing best practice for solid waste management is to prevent pollution. Pollution is a threat to human and other living organism and it may also damage the ecosystem and disrupt the natural cycle and climate on earth [27]. There are many disposal options available to suit the nature of waste and a country's preference and interest. Economics and environmental aspects of waste disposal option are always the main issue in choosing the right technology [28]. Most developed countries, are on their way to eliminate land filling while some other countries still have problems with open dumping [29, 30].

Despite the development of many waste disposal option, landfills remain the most prominent system applied worldwide [30, 31]. Although a lot of improvement had been possible in the land filling system and the regulation on the type of waste that can be treated at landfill is stringent, most of landfills operated remain primitive [31]. Ayomoh et al. [32] had listed few problems related to improper landfill operation including, health deterioration, accidents, flood occurrences, pollution of surface and underground waters, unpleasant odor, pest infestation and gas explosion. Although the impacts from landfills are known, impacts from other alternative remain unanswered thus subject to critics [31]. Incineration has been the choice for developed countries as they have sufficient financial input and are looking into energy recovery from waste [33, 34]. Small countries such as Singapore adopt incineration as their waste disposal option due to scarcity of land [35]. Even that, incineration is also associated with some other risks. This includes the generation of carcinogenic and toxic compound.

Some scholars have observed that the impacts from incineration are overemphasized and the advancing technology had highly reduced the environmental impacts [31]. However, many of the countries prefer waste minimization compared to waste treatment such as landfill or incineration [35, 36]. Technology is advancing every day and chemical recycling of plastic wastes has also been made possible in these developed countries [37]. Regardless of the technology chosen, each has its advantages and disadvantages. The information on each disposal option needs to be clarified to determine the suitable option for each particular country. Few tools had been used in the environmental evaluation including in determining best waste disposal option. For example, life cycle assessment determined that the most economically feasible option for traditional market waste management in Indonesia is composting at a centralized plant, while biogas production option has the lowest environmental impact [28]. SW Plan software particularly to calculate capital and management cost is also available to determine the best integrated technology in waste management [38].

2.1 Solid waste management system in Ghana

Before 1985, incinerators were the technology used for handling waste in the urban centers of Ghana. This could not be sustained due to the lack of funds as a result of economic hardship in early 1980 and technical knowhow. In view of this by 1985 solid waste were dumped on all bola locations [39]. Thereafter a special department called the waste management department (WMD) was set up in the urban centers in 1985 to manage the waste in Ghana with financial and technical assistance from the German Agency for Technical Co-operation (GTZ). The first house to house collection started in Accra using animal drawn carts using donkeys in the high income residential areas. Waste collected was dumped into central containers. Using only 15 donkeys and 10 staff the carriage could collects 3–4 trips daily which covered 75–100 houses [39]. The GTZ project helped to improve the deteriorated waste management in Ghana. However their exit saw more deterioration in level of service quality and service coverage due to the fact that the public provision alone could not handle the growing urbanization of the towns and cities. This however calls for further decentralization to include the private initiative in solid waste management.

The waste companies provided house-to-house and communal services. The communal service was mostly provided in the lower middle income areas using central containers. Residents who patronize this kind of service disposed of their waste by taking it to a central containers site. This containers are lifted full of waste and dispose of at designated disposal sites [39]. Private Sector Initiative (PSI) started in Accra and Tema in the early 1990s and later extended to Kumasi in the mid-1990. Afterwards, this initiative was extended to Takoradi and Tamale in 2000 and 2002 respectively. There year 2004–2007 saw the inclusion of more private companies in to waste business all over Ghana. The companies in Accra and Tema increased to 18 and 6 respectively by 2006. As a result, contracts were open up for competition. The first competitive bidding for solid wastes took place in Kumasi in 2007 and later in Accra in year 2008 [39]. The rapid population growth in Ghana has resulted in increased waste generation in the country. The amount of solid waste generated per day in Accra was 750–800 tonnes in 1994 [40]; 1800 tonnes per day in 2004; 2000 tonnes per day in 2007 this figure increased to 2200 in 2010 [41].

The methods for solid waste disposal in Ghana are uncontrolled dumping of refuse, controlled dumping, sanitary land filling, composting, and incineration [42]. Open refuse dumps are most commonly located at the perimeter of major urban centers in open lots, wetland areas, or next to surface water sources. Open dumps are generally sited based on considerations of access to collection vehicles rather than hydrological or public health considerations. In rural areas and small towns, there are often no vehicles for collection hence uncontrolled dumping occurs within the built up areas with all its attendant health hazards and negative environmental impact [42]. Problems from landfills in Ghana include odor, insufficient covering material, flies and other vermin infestations and smoke from open fires. The increasing amount of waste received by these landfill make it necessary to find other disposal option since constructing new landfills may be difficult due to the scarcity of land, increase of land price and demand for a better disposal system. Effective solid waste therefore calls for a competent and responsible institutions as well as sound managerial system.

The Ministry of Local Government and Rural Development (MLGRD) is the institution responsible for waste management services at the national level. This institution formulates waste and sanitation policies and also provides oversight role to the assemblies and gives subsidies for the provision of SWM services. The Ministry supervises the activities of local Assemblies and passes order as required by law to the various Waste Management Departments of the local Assemblies who are directly responsible for effective solid waste management. As part of the decentralization process in Ghana, in 1988 the waste management functions became a sole responsibility of the Assemblies [43]. About 90% of the Assemblies budget is supported by the Central Government to carry out their obligations in the locality through the various departments. The WMD is responsible for all the waste collection, disposal and monitoring of all the activities of companies engaged by the Assemblies. On the legal and regulatory frameworks for effective solid waste management, the policy which regulates waste management in Ghana is primarily reflective of legislation enacted at the national level and decisions made in pertinent case law. The Central Government bestows local authority status, onto any town or city in accordance with Act 462 which come to replace the previous act enacted in 1988 [39]. In spite of this, the Government continues to exercise controls over the Metropolitan, Municipal, and District Assemblies (MMDAs). The Central Government usually gives directives that affect the Assemblies. The most important is the fact that, a considerable amount of the Assemblies revenue is a direct disbursement from the Central Government. This makes it very difficult for the assemblies to be free from government interference. However, the MMDAs have a constitutional mandate under the 1993 (Act 462) to effectively handle sanitation issue which includes solid-waste management and therefore needed to operate independently to benefit the people. This responsibility is farfetched due to lack of independence. The 1960 (Act 29) of the Criminal Code of Ghana, state in no uncertain terms that whoever places or permits to be placed, any refuse, or rubbish, or any offensive or otherwise unpleasant material, on any yard, street, enclosure, or open space, except for the reason that such a place has been designated by the Assembly for such intent and purpose commits an offense. The law requires individuals to take full responsibility for the streets, drains and space closer to their premises [39].

In addition, the legal regime in Ghana mandate the Assemblies as owners of all the waste generated in municipalities and as a result has the mandate to collect, recycles and discards solid waste. The National Building Regulations, The 1996 (LI 1630) which is the national building regulation stipulates that a building for residential, commercial, industrial, civic or cultural use shall have a facility for refuse disposal, a standardized dustbin and other receptacles approved by the Assembly in which all the waste generated shall be stored pending final collection by the trucks to final disposal site [39]. SWM in Ghana is greatly influenced by the Environmental Sanitation Policy of 2008. This policy is an update of the 1999 policy with the view to meet the prevailing development objectives and address the aspirations of the principal actors in the sector after 8 years of slow implementation with very little impact [10]. With reference to environmental sanitation, the policy requires the Assemblies to control environmental sanitation and check pollution in all forms [39]. The policies tend to reflect prevailing ideas on solid waste management and give an overall evaluation of the prevailing circumstance in the country. It further ensured private sector participation and the provision of 80% of SWM in all the assemblies [39]. The Ministry of Local Government is mandated to regulate the waste business. The regulation works to promote competition via legal restrictions and regulatory rules and controls concerning market entry and exit, the capacities of companies operating in the waste market, user charges and the service standards. The local assemblies are mandated to outsource solid waste collection to decentralized agents service by contracts and also embark on frequent monitoring and evaluation of the service quality provided by the companies and sanction any insubordination according to the dictate of the contract.

The policies and regulations and the contractual agreement that connect the assemblies with the companies are important factors that contribute to effective solid waste collection, treatment and disposal. These regulations include the Local Government Act, National Procurement Act, Local Governments By-law, Environmental Sanitation Policy, and other state conventions that provide rules for solid waste management. The Procurement Act [44] requires the Assemblies Tender Boards to use competitive bidding to select companies [39]. This call for appropriate mechanisms suitable for the local conditions from an environment, social and fiscal perspectives, and at the same time being more capable to be sustain over long period of time without reducing the resources it needs [45]. Based on this the conceptual framework of the study focuses on four key variables, namely: evolving practice of SWM, households' involvement for service sustainability, private company capacity and lastly, regulatory mechanisms and control for solid waste management in relation to service quality.

3. Methodology and study setting

3.1 Study setting

This study focused on two municipal areas (Berekum and Dormaa) located in the Brong Ahafo region of Ghana (see **Figure 1**). These municipalities were selected based on their rapid expansion and urbanization [46]. The total land size of the Berekum Municipality is 1635 km^2. This area covers about 0.7% of the entire land area of Ghana (233,588 km^2). The Berekum Municipality lies between latitudes 6° 27 N and 7°00 N and longitude 2°52 W. According to the 2010 Population Census of Ghana, the population of the municipality stood at 129,628. The annual average population growth rate is 2.2%. The 2015 population of the municipality was 144,528. This growth rate compares favorably with both the regional and national rates of 2.3 and 2.5% respectively. Dormaa Municipality, on the other hand, lies between latitude 7° and 7°30'N and longitude 3° and 3°30'N. It covers a land area of

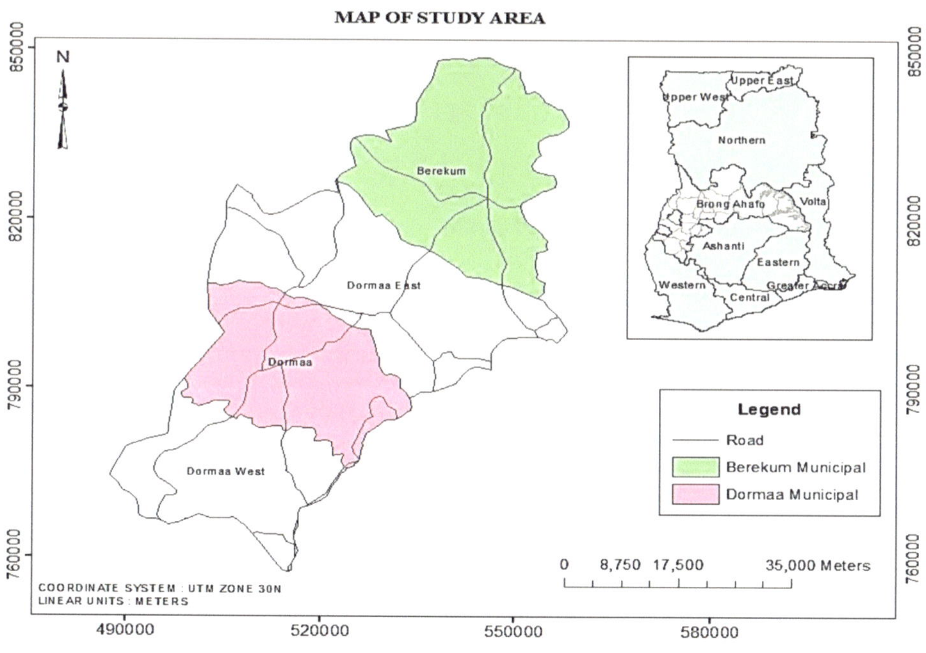

Figure 1.
Map of study area.

912 km^2. The 2010 housing and population census of Ghana put the total population of Dormaa municipality at 159,789 with an annual growth rate of 2.4%.

3.2 Research design

The study adopted the case study research method [47]. Purposive sampling [48] was used to select 12 communities from the two municipalities. Firstly, the study area was zoned into two clusters namely: Berekum municipality and Dormaa municipality. Secondly, purposive sampling was used to select twelve (12) areas from the two municipalities for the survey. Through a mixed methods design [49], both qualitative and quantitative research methods were used for the data collection and analysis. A household survey was conducted with household respondents to understand solid waste management and service delivery across the 12 selected communities. Using Slovin's formula: n = N/1 + N(α)2, where 'n' is the sample size, 'N' is the total number of households, 'α' is the margin of error (0.05), a total sample size of 312 households across the 12 case study communities were randomly selected and involved in the household survey. The sample size of 312 was divided equally among the 12 selected communities. This gave a sample size of 26 for each selected area. Finally, accidental sampling method was used to select the respondents for interview. That is, the first person to be contacted in each selected house was interviewed. If the first person contacted was not ready, the next available person was interviewed. To gather statistical and policy information on solid waste management and service delivery the two municipalities, semi-structured interviews were also undertaken with an official of the Assemblies (Berekum and Dormaa), responsible for the environmental health and waste management of the municipalities. In analysis, the study used a cross-case analysis procedure to analyze the interview data. In this approach, responses to a common question from all interviewees in each category are analyzed together. The findings of the study were validated and verified through focus group discussions with household respondents in each of the 12 case study communities. This approach was appropriate in addressing the inconsistencies that had occurred during the data analysis.

4. Results and discussion

4.1 Evolving trend of decentralized SWM

Coverage of service in these two municipalities as a result of GTZ assistance in the 1980s were not available it is believe that coverage were very high. This is seen in the numerous waste dumping site which became known as "bola" in the old communities of these two municipalities. As these municipalities expanded the waste departments did not build new site for waste. With this, the coverage continued to fall from the 1990s of about 75–50 perfect by the year 2006 according to the municipal waste directors of Berekum and Dormaa. The fall off called for the involvement of the private companies. Checks by the study revealed that the private companies formally started in Berekum and Dormaa in 2006. These municipalities were not group into zones. One company provided SWM services in the municipalities, there was no competitive bidding, and one company was given the contract to provide house to house and community collection service. This discovery confirms what Oduro-Kwarteng [39] indicated of the evolution of decentralized SWM in the country.

DOI: http://dx.doi.org/10.5772/intechopen.81894

4.2 Modes of waste disposal in Berekum and Dormaa

Basically results from the data analyzed on waste disposal in the Berekum Municipality confirmed that three ways of waste collection exist in Berekum including House-to-house collection, communal dumpsites and open dump site (**Table 1**). From the survey, majority (62%) of the respondents disposed of their waste into communal containers. This is followed by 29% of the respondents who indicated that their waste was collected directly from their houses (which are mostly found in the new residential areas). A total of 16% indicated that they emptied their waste into open dump sites. Similar responses were observed in the Dormaa Municipality as most (63%) of the respondents indicated that waste was disposed of into communal containers. While 20% of the respondent said that waste was collected directly from their house, 17% also indicated that they emptied their waste into open dam sites.

In the Berekum Municipality, the data analyzed shows that the House-to-house service of refuse disposal is primarily practiced in the new residential areas including Nyamenae and Awerempe-Estate. Similar results showed the same trend in the Dormaa Municipality as residential areas such as Kumidaa Street and Asikafo Amantem were found to be practicing house-to-house waste collection. These modes of waste collection were verified with key stakeholders (the Assemblymen, WMD and Private waste company). The introduction of this service in the municipalities reflect the trending urban form of solid waste management since such areas compose of settlements which house middle to high income earners who are in the position to pay for such service. As Oduro-Kwarteng [39] asserted in the formal introduction of this service in the urbanized areas in the two major cities in Ghana (Accra and Kumasi). Other towns and cities have grabbed this concept to enhance service delivery as far as SWM is concerned. All the respondents from these residential areas where house-to-house waste collection service takes place in the Berekum Municipality are required to pay a monthly charge of GH¢15 (US$3) per 120 liter dustbin. In the Dormaa Municipality, service beneficiaries pay an amount of GH¢10 (US$2) per 120 liter dustbin. The results show that service beneficiaries in the Dormaa Municipality slightly pay lower price than amount paid in the Berekum Municipality. According to the Assemblymen this charge was exorbitant and as a result accounted for the lack of patronage in the Municipality. Secondly, there was lack of patronage because the companies did not regularly and routinely collect waste in these areas. In view of this some people turned to burning as a means of dealing with their waste.

Communal collection was mainly carried out in the old town residential areas of Kyiritwede Zongo and Amangoase for Berekum and Atoase, Ahantrase Ahenbronofor Dormaa Municipal. This mode of waste collection does not require

	Do you pay for collection service?						
Berekum	**Yes**	**No**	**Total**	**Dormaa**	**Yes**	**No**	**Total**
Door to door	41	0	41		31	0	31
Communal collection	0	97	97		0	98	98
An open dump	0	18	18		0	27	27
Total	27	91	156		23	85	156

Source: Field Survey, 2015.

Table 1.
Modes of waste disposal in Berekum and Dormaa.

any monthly fee or pay as you dump charges. Residents go to a central container and dispose of their waste. The next mode of collection is the open dump site collection this is seen in the suburbs; communities, who dispose of their waste at the open dump site, are emerging communities that the Assembly together with private companies have failed to supply with containers. In view of this, the people throw their waste in open dump pit this is seen in some part of Atonotia and the light industrial area of Berekum municipal and in New Dormaa for Dormaa Municipal. This findings supported studies by [42] who asserted that open refuse dumps are most commonly located at the perimeter of major urban centers in open lots, and are generally sited based on considerations of access to collection vehicles.

4.3 Households involvement in solid SWM in Berekum and Dormaa

This study also examined the extent to which the various households in the two municipalities participate in waste management services in relation to the mechanism for cost recovery, the eagerness-to-pay for service charge, eagerness to separate waste at source and monitoring of service quality. **Table 2** presents the household's views on the assessment on who ought to bear the cost of waste collection in the municipality. The study shows that, 58% of the respondents within the two municipalities who utilize the house-to-house waste service perceived both the Assembly and the individuals who generates the waste have to work very hard to recover more than 50% of the cost incurred if not all in waste collection and disposal. On the other hand, 24% opted for the generators to incur all the cost involved in waste management without any prejudice. Whereas 18% said the Assembly alone should incur the cost for waste management services.

Regarding communal collection about 54% of the respondents said the Assembly alone should pay for the cost of waste collection, while about 45% indicated that the generator and the Assemblies have to collectively pay for waste services. Moreover, only 5% said only generators should pay for waste services. In view of the above, it is quite obvious that the companies need to be more responsible for results and to be more responsive to their client. This also implies that much attention must be given

	N	Berekum	Dormaa	%
House-to-house	**72**	**41**	**31**	
Generator only	17	10	7	24
Generator & Assembly	42	18	24	58
Assembly only	13	13		18
Communal collection	**195**	**97**	**98**	
Generator only	10	7	3	5
Generator & Assembly	87	34	53	45
Assembly only	98	56	42	50
Open dump	**45**	**18**	**27**	
Generator only	2	0	2	5
Generator & Assembly	14	14	0	31
Assembly only	29			64
Source: Field Survey, 2015.				

Table 2.
Opinion on who ought to bear the cost of waste collection services.

to household involvement to make sure the households are well informed about the fiscal problem confronting the Municipalities and the necessity to pay for service improvement.

4.4 Eagerness to pay for service charge

To further ascertain the household involvement in solid waste management, the resident's eagerness to pay more for waste services was assessed. **Table 3** indicates the results of the eagerness-to-pay service charges. To improve the effectiveness of the house-to-house service, the respondents were ask on their willingness to-pay more. The result on this shows that a total of 21% of the people interviewed were willing to pay more for waste services. This was due to the fact that the respondents were not satisfied with the existing service quality. However, the majority (53%) of the respondents confirmed their eagerness to pay the existing tariff for the service rendered, whereas a total of 26% were eager to pay less than prevailing tariff. This group saw the service quality to be very poor and that, wanted an improvement in service quality levels before tariffs are increased. This finding support what many call the need for government to encourage the principle of polluter-pays which financially resource service providers in service delivery.

The result further indicated that, all the respondents were ready to pay any considerable tariff for the service if the service would be improved along frequent and routine waste collection of two times a week. It was again realized that the respondents wanted the tariffs to be charged on waste volumes and rate with which waste is being picked up. Regarding communal service, it came out from the study that all the respondents did not pay for services. However, over 65% from the survey were eager to pay for the tariff under one condition that service improves. The implication is that more effort should be geared toward educating the public and for that matter the customers to come to terms with the need to pay for services to recover cost to ensure better service quality. Moreover, the companies ought to be more responsive to complains of the customers so as to improve service quality.

Households eagerness to pay for services	Berekum	Dormaa	Total	%
House-to-house collection	**41**	**31**	72	
Eagerness to pay more	9	6	15	21
Eagerness to pay current user fees	21	17	38	53
Eagerness to pay less than the current user fees	11	8	19	26
Communal collection	**97**	**98**	**195**	
Pay tariffs at time of survey	0	0	0	
Eagerness to pay for the service	76	63	139	71
Open dump	**18**	27	**45**	
Pay tariff at time of survey	0	0	0	
Eagerness to pay for the service	12	10	22	49

Source: Field Survey, 2015.

Table 3.
Respondents' eagerness-to-pay user charges.

4.5 Involvement of households' in waste minimization

On the residents' readiness to separate and recyclable their waste at the house (source) for collection, the result indicates that majority (50 and 49%) of the respondents in Berekum and Dormaa respectively were ready to separate their waste at the source given the necessary incentives. They pointed the increase in collection rate to two times a week, the free provision of plastic bags with variety of colors, and to be provided with free bins by the companies or the assembly for separate collection as the incentives needed for effective waste separation. Over 40% accepted to purchase their own receptacles for storing organic waste. Whereas 33% called the enforcement of by-laws to ensure everybody separate their waste. The respondents acknowledged their awareness on waste reuse, recycling, as well as composting. Majority indicated that they use food waste to feed livestock, salvage used plastics and cans, and sachet rubbers for the informal buyers or scavengers. Moreover some continue to engage in burning waste. Small number of them uses organic waste as manure for vegetable garden. Notwithstanding, the residents' awareness on reuse and recycling of waste in the municipalities are very low as more reusable and recyclable materials continue to be seen in the streets, drains and streams. This implies that the existing collection system does not ensure recycling as varied wastes are sent to the dumping grounds with very little or no recycling by scavengers.

4.6 Service quality of waste management in Berekum and Dormaa

The quality of SWM was assessed by asking the respondents to indicate service satisfaction by responding either satisfied or not satisfied with the quality of service on a five-point scale from very poor to very good in terms of two service quality attributes (reliability of collection and sanitary conditions at bin/container location). To rate the quality SWM of the service providers effectively, all the communities served by the waste management company were selected for the survey. The study shows considerable disparities in terms of quality in the existing SWM system as practiced in the municipalities. In the Berekum Municipality, a total of 65% of the respondents who patronize house-to-house collection service rated there liability of service and sanitary condition and waste overflow as fair and good. Areas such as Estate, Nyamebekyere and Osofokyere which have larger number of high and medium income households' fall in this category of the respondents rated the quality of service of the company in their vicinity as good. The level of service quality could be attributed to the perceived quality of service by the people. This is because they pay for waste services that recover full cost and therefore they expect the service to be devoid of waste overflow from bins located in front of their house. The service reliability and sanitary conditions of communal collection in the low income areas of Atonotia, Kyirikwede and Amangoase were largely rated as poor by residents. Surprisingly, a total 76% of the respondents for communal collection rated the service as poor. This is because the waste overflow from communal containers unto the ground was widespread. The people in these areas confirmed that collection is irregular and the containers 'sites are not desirably maintained by the company and the Assembly.

In the Dormaa Municipality, the survey revealed similar results. There were also considerable disparities in terms of quality in the existing SWM system as practiced in the municipality. About 71% of the respondents who patronize house-to-house collection service rated the sanitary condition and waste spill over as good. In addition 59% said the reliability of waste collection was also good. Areas such as Kumidaa Street and Asikafo Amantem which have larger number of high and medium income households' fall in this category of the respondents rated the quality of service of the company in their vicinity as good. The service reliability and sanitary conditions of communal collection in the low income areas of Atoase, Ahantrase and Ahenbrono were generally rated as poor by residents. Interestingly, about 67 and 74% in these areas rated the sanitary condition and Reliability of waste collection respectively as poor. To them, the rate of waste overflow, from communal containers unto the ground at the container sites were high. The households confirm that collection is irregular and the containers 'sites are not cleaned by the company. Comparatively, more of the residents in Dormaa Municipality rated the service quality for the house to house as good than those from Berekum Municipality. Approximately 71 and 59% of the residents from Dormaa Municipality rated the sanitary condition at the container site and the reliability of waste collection respectively as good whereas 65 and 54% also rated the sanitary condition at the container site and the reliability of waste collection respectively in Berekum Municipality. Similar, results came out regarding communal collection. A total of 43 and 36% rated the sanitary condition at the container site and the reliability of waste collection respectively as good from the Dormaa Municipality whereas 27 and 24% also rated the sanitary condition at the container site and the reliability of waste collection respectively in Berekum Municipality.

4.7 Mechanisms for solid waste management regulation

To monitor the quality of service effectively, the companies are mandated to furnish the Local Assembly with information on monthly basis. This comprises of performance targets, vehicle tour schedule, proceeds and expenditure from house-to-house collection and tonnage of waste disposed of. The key informants revealed that the performance targets as well as the formal rules and regulation for private waste companies were obviously elucidated in the contract signed. In addition, they affirmed that, the company cooperate with the Municipalities and provide information on tonnage on waste collected. This information is kept and used as the basis for paying the companies. With reference to house-to-house service, the companies further admitted that they (companies) provide the municipalities with information on revenues from the house-to-house services. In contrast, the staffs of the WMD were of the view that actual revenue from house to house collection is not properly accounted for in the reported to the Assemblies. Further result from them pointed out that detailed document on claims and revenue collected always lag behind time and the revenue figures usually were far below expectation. There was a clear evident of information asymmetry with the reports on cost and revenue in all the two municipalities. The information asymmetry in the report of the companies did not arguer well for the Assemblies to have a firm grip on cost and revenue to make any meaningful plan for effective cost recovery mechanism.

The results further revealed that the Assembly alone set up the service charge for the communal as well as house-to-house service. They further pointed out that the tariff for house-to-house services devoid of any central government support are fixed by the individual companies and submitted to the Municipal Assemblies for approval. The Assemblies specify an indicative levy for house-to-house collection to be collected monthly and a unit price per emptying the skips for communal collection service as specified in the contract document. The final levies and the unit prices at are susceptible to changes using the price escalation formula in the contract after the award of contract. Concerning waste collection charges and fee (unit price) and cost recovery, the companies providing the house-to-house collection takes approved service fees from their client on monthly basis in both Berekum and Dormaa. The house-to-house collection fees for waste management were GH¢12

(US$2.50) Berekum and GH¢10 (US$2) for Dormaa but those who patronize the communal services do not pay for user charges. However the cost for lifting a tonne of waste keep increasing with time and over time, this has become a burden on the assemblies. It was further revealed that the user charges were not regularly reviewed. This has resulted in big cash flow problems for the companies due to the continuous increase in exchange rate of the cedi, inflation and fuel prices. The user charges need to be reviewed by the Assembly and published the new fees in national gazette as by-law for it to be legally binding on residents. This according to Assemblies is cumbersome and requires political will on the part of the central government and municipalities. This indicates clearly that there is a look warm attitude from the Municipal Authorities to implement full cost recovery through charging of all households in the two municipalities.

Also, the key informants revealed that the Assemblies hardly conduct public education. There was only two and four count for Berekum and Dormaa Municipalities respectively. This has adversely affected the residents' attitude toward waste management. The residents continue to litter indiscriminately. In relation to the Assembly's commitment to bye-laws it revealed that the two assemblies had bye-laws fully gazetted to keep the companies and the residents within the confines of SWM best practices. However, the bye-laws were not strictly enforced. Subsequent result shows that the Assembly finds it difficult to fulfill the terms of payment as stated in SWM contractual arrangements with private companies. More so, there had not been any occasion where interest had been paid on delayed payments beyond the 3 months as stated in most contracts. The difficulties and holdups identified in the Assemblies commitment to its contractual obligations are basically lack of financial resources. The key informant from the waste management department said they keep on changing their schedule for educating the masses on waste management year in year out all because of the lack of funds. Conversely, according to the companies, the cost recovery mechanism is inadequate. With this the assemblies find it difficult to generate enough revenue to pay the companies. There was also weak mechanism in place to deal with residents who refuse to pay for the waste collection services rendered. The Assembly delays so much with the payment of monies and this in effect affects service quality. It can be concluded that Assemblies' non-adherence to contract obligations have a major influence on service quality and productivity of companies.

Also, concerning the companies' commitment to contractual obligations, indicators used includes company achieves daily collection target in the contract, company's cover waste containers during transporting, company collection crew use protective clothing, company keeps container site free of litters and clean. The directors of both the private companies and the WMD interviewed said the companies were able to achieve their daily targets of about 80%. It was confirmed together with other key informant and the resident in the household survey that the collection crew have protective clothing and use them their activities, however very few about 10% refuse to wear theirs in most cases. The few workers who do not in most cases use the protective clothing started with the informal sector and believed they are responsive to the waste collection without protective clothing. Regarding the companies obligation to keeping the container sites clean especially with the communal collection. Twenty-five out of over 100 container sites were kept clean and tidy whiles the others had litters all over. Large heaps of waste remains at these container sites after solid waste has been move to the disposal sites. In relation to this is the companies' obligation to repair and maintain communal waste skips. The key informant said the company barely does this function it is only the assemblies that squeeze some funds out of pressure from the residents for few repairs works on these containers.

Regarding the enforcement of legislation and sanctions, the municipal assemblies have the sole responsibility to enforce legislations and sanctions on the provision of public services. The Assembly uses bye-laws as well as terms and condition in the contract as the basic mechanism to managing solid waste collection services in their area of jurisdiction. It was realized from the companies' point of view that the bye-laws were enforced. In addition they were also of the view that the monitoring of compliance was done effectively. In addition, they opined that the sanctions for noncompliance to the bye-laws were punitive enough. They also revealed that the environmental health standards and sanitation were strictly observed and enforced. However, the household survey shows a different picture. It indicated that very little have been done to enforce bye-laws. It was realized that, the Assemblies find it very difficult to sanction offenders due to the frequency at which these bye-laws are flouted. The residents show lax attitude toward effective waste management. It was also observed that there were inadequate waste containers and low frequency of waste collection especially with the communal collection. Relating to this is the lack of environmental sanitation courts in these areas. This hinders the enforcement of solid waste and sanitation regulation.

The fines for non-compliance are the same in the two municipalities and are subject to review. They may be changed by the Assemblies after the service provider has been informed of such changes. The fine ranges from GH¢100 (US$20)-GH¢200 (US$40). The study revealed that the municipal assemblies have so far not been able to apply any sanction to the companies though evidence from the household survey shows the companies fail enormously in waste pick up as well as the cleaning up of the container sites. This study is therefore consistent with the finding of Oduro-Kwarteng that there is lack of sanctions in the waste management sector. In a similar study involving five cities, Oduro-Kwarteng [39] discovered that many of the contracts had credible threats of sanctions that required sanction, but non-complying companies were not penalized.

5. Conclusion and recommendations

5.1 Conclusion

The study focused on the evolving SWM practices, the quality of service as well as the factors that influence the private sector performance and their implications for solid waste collection in the medium towns. The study revealed that there were no significant disparities in service quality among the two municipalities. But more difference do exists among different communities due to the difference in the methods of waste collection. The service quality of house-to-house collection practiced at well-organized residential areas was higher than that of communal collection at old town lower income residential areas. The study further revealed that more waste is now been collected than before due to increasing role of the private sector in the waste business. Over 80% of waste generated in these municipalities are collected and send to a designated site for final disposal by the private sector waste management firms. This is much better than the 2006 figure of about 50–80% waste collection. In addition, the participation and involvement of households at any level of the SWM has been very slow principally due to the lack of funds and public education. Moreover, a shift toward cost recovery through charging all households a fixed charge for house-to-house collection is in places. However there were no mechanisms for full cost recovery to include majority of the residents who patronize communal collection service. The assembly therefore faces problems of financing bins, providing for other resources which resulted in to illegal dumping by some households.

Furthermore, the study revealed that there were weak regulatory practices and non-adherence to contractual obligations and these consequently provided no incentives for full cost recovery and better service quality. The weak regulatory practices such as no competitive bidding, prolonged periods before upward review of collection fees and service charges, no interest on monies delayed, and delay in payment of subsidy does not provide incentive mechanism for private sector growth and does not enhance better waste management service delivery. Again, the study revealed that there is a weak institutional capacity (inadequate personnel and logistics). The responsibility over solid waste collection and disposal is well beyond the capacity of waste management institutions. They could not collect the 20–80% waste as stated in the contract document due to lack of personnel and logistics for monitoring and supervision. However, it was revealed that the involvement of the private companies in the management of solid waste has strengthened the capacity of the municipal assemblies. Yet, this is still deemed inadequate to meet the required levels of urban solid waste collection. Finally, the study revealed that there is a lack of strict monitoring and enforcement of sanitation bye-laws in the municipalities. However, the study found that the bye-laws were punitive enough but lacked strict enforcement. The non-enforcement of the bye-laws has contributed immensely to the indiscriminate dumping in the municipalities.

5.2 Recommendations: emerging interventions

In order to address the problem of municipal solid waste in the study areas in particular and Ghana in general, it is strongly recommended that the policy mechanisms and strategies adopted should be holistic and comprehensive. The nature of the issues and challenges identified require multidimensional interventions in order to provide sustainable solutions. There is the need for clearly defined standards and service quality in the contract for regulating the private sector activities. This will facilitate a well-managed SWM system in the municipalities as the private companies were not abreast with these standards and the terms of the waste management contract. The study also recommends a capacity building training on waste management for the officials of Waste Management Department in the municipalities as well as the technical operation officers of the private sector waste management companies.

There should also be full cost recovery for waste services. This requires the 'pay as you throw' (PAYT) mechanism for communal collection to ensure financial sustainability and quality service delivery. Although such mechanism has failed at initial stage in Accra in 1995, it worked well in Kumasi. The success of the PAYT in Kumasi was due to the participation and creating of public awareness, household participation and involvement at all levels as well as the enforcement of bye-laws on indiscriminate dumping. The prevailing system where communal collection is free for resident is not sustainable. Also, Assemblies should to be encouraged to be responsive to effective and quality service delivery. The environmental health unit should be restructured to make it more responsive to the challenges of SWM. The environmental health personnel can also be attached to the private companies to enforce bye-laws on paying for service and prevention of indiscriminate dumping of waste. Furthermore, the establishment of recycle firms should be encouraged by the Assemblies. They can start by forging partnership with the private sector companies. Also, the coordination for waste management should be encouraged within the context of environmental education and stricter enforcement of sanitation bye-laws. This is because environmental education creates environmental awareness and makes people conscious of environmental and sanitation issues. The enforcement of bye-laws is important in view of the fact that environmental awareness is not sufficient enough to ensure change in behavior. Therefore, stricter law enforcement is needed to deter people from dumping indiscriminately.

Author details

Richard Kyere[1], Michael Addaney[2,3]* and Jonas Ayaribilla Akudugu[4]

1 Ghana Education Service, PMB, Kumasi, Ghana

2 Research Institute of Environmental Law, Wuhan University, Wuhan, China

3 University of Energy and Natural Resources, Sunyani, Ghana

4 Department of Community Development, Faculty of Integrated Development Studies, University for Development Studies, Ghana

*Address all correspondence to: appl.adm@gmail.com

References

[1] UNDESA/PD. World Urbanisation Prospects: The 2011 Revision. New York: United Nations; 2012

[2] Cobbinah PB, Addaney M, Agyeman KO. Locating the role of urbanites in solid waste management in Ghana. Environmental Development. 2017;**24**: 9-21. DOI: 10.1016/j.envdev.2017.06.004

[3] Cobbinah PB, Aboagye HN. A Ghanaian twist to urban sprawl. Land Use Policy. 2017;**61**:231-241. DOI: 10.1016/j.landusepol.2016.10.047

[4] Addaney M, Oppong RA. Critical issues of municipal solid waste management in Ghana. Journal of Energy and Natural Resources. 2015; **2**(1):30-36

[5] Litvak J, Juniad A, Bird R. Rethinking Decentralization in Developing Countries. Sector Studies Series. Washington, DC: The World Bank; 1998

[6] United Nation Environmental Programme (UNEP). International Source Book on Environmentally Sound Technologies for Municipal Solid Waste Management. USA: UNEP; 2000

[7] World Bank. Facts and Figures, Environmental Dal Tables. 04.03.2000; 2007

[8] World Bank. Ghana Country Environmental Analysis. Report No. 36985-GH, 2.11.2007; 2007

[9] Hoornweg D, Bhada-Tata P. What a Waste: A Global Review of Solid Waste Management. Washington D.C.: The World Bank, Urban Development Series 15. 2012. pp. 8-10

[10] Ministry of Local Government and Rural Development (MLGRD). Environmental Health and Sanitation Directorate. In: National Environmental Sanitation Strategy and Action Plan (NESSAP). 2010. pp. 2010-2015

[11] Contreau-Levine S, Adrian C. Guidance Pack: Private Sector Participation in Municipal Solid Waste Management. St. Gallen, Switzerland: Swiss Center for Development and Corporation in Technology and Management; 2000

[12] World Bank. The Challenge of Urban Government: Policies and Practices. Washington: The World Bank Institute (WBI) Development Studies; 2000

[13] Bilitewski B, Härdtle G, Marek K. In: Weisegch A, Boeddicker H, editors. Waste Management: Abfallwirtschaft: eine Einführung. Berlin/New York: Springer; 1994

[14] Pongrácz E. Re-defining the Concepts: Re-Defining the Concepts of Waste and Waste Management: Evolving the Theory of Waste Management. 2014. ISBN 951-42-6821-0

[15] United Nations Environment Programme (UNEP). Solid Waste Management. Vol. I. USA: UNEP; 2005

[16] Pfeffer JT. Solid Waste Management Engineering. Enlewood, NJ: Prentice; 1992

[17] Dorvil LP. Private sector participation in integrated sustainable solid waste management in low- and middle income countries [PhD thesis]. The University of St Gallen, Graduate School of Business Administration, Economics, Law and Social Sciences (HSG); 2007

[18] Watchers W. Solid Waste Definition (Printable version). Medicine Hat: Alachua Exchange; 2004

[19] Tchobanoglous G, Theisen H, Vigil S. Integrated Waste Management: Engineering Principles and Management Issues. New York: McGraw-Hill; 1993

[20] Hazra T, Goel S. Solid waste management in Kolkata, India: Practices and challenges. Waste Management. 2009;**29**:470-478

[21] Wagner T, Arnold P. A new model for solid waste management: An analysis of the Nova Scotia MSW strategy. Journal of Cleaner Production. 2008; **16**(4):410-421

[22] Puopiel F. Solid waste management in Ghana: The case of Tamale Metropolitan Area [MSc thesis]. Kumasi, Ghana: Kwame Nkrumah University of Science and Technology; 2010

[23] Kathiravale S, Muhd Yunus MN. Waste to wealth. Asia Europe Journal. 2008;**6**(2):359-371

[24] Panel on Climate Change (IPCC). Waste Generation Composition and Management Data: Guideline for National Greenhouse Gas Inventory. Paris: IPCC/OECD/IEA; 2006

[25] Schübeler P, Wehrle K, Jürg C. Conceptual Framework: Conceptual Frame work for Municipal Solid Waste Management in Low-Income Countries: Working Paper No. 9. St Gallen: SKAT; 1996

[26] Mantell CL. Waste Management. London: John Wiley and Sons Inc; 1972

[27] Raga GB, Castro T, Baumgardner D. The impact of megacity pollution on local climate and implication for the regional environment: Mexico City. Atmospheric Environment. 2001; **35**(10):1805-1811

[28] Aye L, Widjaya ER. Environmental and economic analyses of waste disposal options for traditional markets in Indonesia. Waste Management. 2006; **26**(10):1180-1191

[29] Agamuthu P, Fauziah SH, Kahlil K. Evolution of solid waste management in Malaysia: Impacts and implications of the solid waste bill, 2007. Journal of Material Cycles & Waste Management. 2009;**11**:96-103

[30] Shekdar AV. Sustainable solid waste management: An integrated approach for Asian countries. Waste Management. 2009;**29**:1438-1448

[31] Hamer G. Solid waste treatment and disposal: Effects on public health and environmental safety. Biotechnology Advances. 2003;**22**:71-79

[32] Ayomoh MKO, Oke SA, Adedeji WO, Owaba OEC. An approach to tackling environmental and health impacts of municipal solid waste disposal in developing countries. Journal of Environmental Management. 2008; **88**(1):108-114

[33] Kleiss T, Imura H. The Japanese private finance initiative and its application in the municipal solid waste management sector. International Journal of Project Management. 2006; **24**(7):614-621

[34] Papageorgiou A, Barton JR, Karagiannidis A. Assessment of the greenhouse effect impact of technologies used for energy recovery from municipal waste: A case for England. Journal of Environmental Management. 2009;**90**(10):2999-3012

[35] Bai R, Sutanto M. The practice and challenges of solid waste management in Singapore. Waste Management. 2002; **22**(5):557-567

[36] Boyle CA. Solid waste management in New Zealand. Waste Management. 2000;**20**:517-526

[37] Al-Salem SM, Lettieri P, Baeyens J. Recycling and recovery routes of plastic solid waste (PSW): A review. Waste Management. 2009;**29**(10):2525-2643

[38] Fauziah SH, Agamuthu P. SWPlan software application for Malaysian municipal solid waste management. Malaysian Journal of Science. 2007; **26**(1):17-22

[39] Oduro-Kwarteng S. Private sector involvement in urban solid waste collection [PhD thesis]. Rotterdam: Erasmus University; 2010

[40] Asomani-Boateng R. Closing the loop: Community-based organic solid waste recycling, urban gardening, and land use planning in Ghana, West Africa. Journal of Planning Education and Research. 2007;**27**(2):132-145

[41] Oteng-Ababio M. E-waste: An emerging challenge for solid waste management in Ghana. International Development Planning Review. 2010; **32**(2):191-206

[42] Danso KBM. The nature of solid waste management in Ghana: Towards data collection for good management practices. 2011. http://www.google.com/webhp?hl=en#hl=en&q=The+Nature+of+Solid+Waste+Management+ [Accessed: 10 August, 2016]

[43] Adarkwa KK. The reality of decentralisation implementation in Ghana and some emerging concerns. In: Egziabher TG, van Dijk MP, editors. Issues and Challenges in Local and Regional Development. Addis Ababa; 2005

[44] Republic of Ghana. Public Procurement Act (663). Accra: Parliament of Ghana; 2003

[45] van de Klundert A, Anschütz J. Integrated Sustainable Waste Management—The Concept. Tools for Decision-makers Experiences from the Urban Waste Expertise Programme (1995–2001); 2001

[46] Ghana Statistical Service (GSS). 2010 Population and Housing Census. Summary Report of Final Results. Accra, Ghana: Sakoa Press Limited; 2012

[47] Yin RK. Case Study Research: Design and Methods. 4th ed. London: Sage; 2009

[48] Guarte JM, Barrios EB. Estimation under purposive sampling. Communications in Statistics-Simulation and Computation. 2006; **35**(2):277-284

[49] Driscoll DL, Appiah-Yeboah A, Salib P, Rupert DJ. Merging qualitative and quantitative data in mixed methods research: How to and why not. Ecological and Environmental Anthropology. 2007;**3**(1):19-28

Chapter 9

Household Willingness to Pay for Improved Solid Waste Management Services: Using Contingent Valuation Analysis in India

Muniyandi Balasubramanian

Abstract

Solid waste management is one of the crucial problems in India. An increasing population, industrialization and urbanization have major sources for increasing solid waste in India. The per capita waste generation in India is between 0.6 and 1 kg per day also expected to increase in future. This chapter has discussed two important aspects first; there is lack of study on economic analysis India, second most of the studies have focused on urban solid waste management in India. The present study has used household willingness to pay through the contingent valuation method for improved solid waste management of 150 household in semi-urban areas in Madurai, India. The study has found that the household respondents are willingness to pay Rs 24 (US$ 0.34) for clean environment in the semi-urban area. This study has also found more than 95% of household respondents are willing to pay for solid waste management in Madurai. Most of the household respondents are felt improper solid waste management has one of the important reasons for health issue particularly for children and elderly people in the study area. The main policy implication of the study is to design proper solid waste management plan for collection, transportation, disposal and segregation of solid waste in semi-urban areas in India.

Keywords: solid waste, willingness to pay (WTP), recycling, India

1. Introduction

Solid waste management (SWM) is continuous to be a major challenge in developing world. Due to lack of appropriated planning inadequate governance, resource constraint ad ineffective management, solid waste especially insufficient collection and improper disposal of it is major problem for developing countries [1–10]. Solid waste in developing countries are less generated compared developed countries (see [2, 11, 12]. Solid waste generation is an increasing global environmental problem [13]. Moreover, most of the developing countries are still in the early stage of their urbanization and economic development process, people generally believe that a fast increase in solid waste generation should be unavoidable in the developing countries [12]. Solid waste collection is one of the important problems in developing world like India. Smaller cities and town collect less than 50% of solid waste per day. Poor solid waste collection has creating many environmental and health problem in city in general particular in semi-urban areas. The annual waste generation has been observed to increase in proportion to the rise in population and urbanization, and issues related to disposal have become challenging as more land is needed for the ultimate disposal of these solid waste [14] More recently, cities have begun paying more attention to enhancing municipal system and suitable solid waste service delivery with special emphasis on involving the private sector.

Poor solid waste management in the developing countries consists of a major threat public health and environmental quality and reduces the quality of life particularly for the poorer residents in both urban and rural areas [12]. This paper reveals the supply side of solid waste management services have always been the major environmental problem in India previous research did only reduce the waste quantities and increasing recycling, landfilling, generation, collection and economic analysis if so, especially contingent valuation method for willingness to pay by how much? Although many studies have been carried out to answer this question.

1.1 Solid waste problem in India

India is the second largest nation in the world with a population of 1.21 billion, accounting for nearly 18% of world's human population, but it does not have enough resources or adequate systems in place to treat its solid waste. Its urban population grew at rate of 31.16% during the last decade to 377 million, which is greater than the entire population of the United States, the third largest country in the world accounting to population [15]. Solid waste management is a significant and growing problem in many urban areas in India due to economic development, urbanization, and improving living standard in cities of developing India have led to increase in the quantity of complex composition of municipal solid waste. Management of municipal solid waste resulting from rapid urbanization has become a serious concern for government departments, pollution control agencies, and regulatory bodies and public in most of the cities in India. The challenges of solid waste in Indian cities and town it addressed by various agencies the responsibility of the collection, removal and disposal of garbage from public places in urban areas and maintenance of dumping ground however, comes under the purview of the local municipal body which is the main formal stakeholder involved in the governance of solid waste management in India [16]. Solid waste management has been the most neglected area of urban development over the years and has accounted for severe health problems in urban areas all over the country. A number of cases have come to light because of mismanagement of municipal solid waste management [17]. Solid waste management has been major concern in developing India see [18–23] in urban areas. Moreover, increasing consumerism and development of technology also has increase in solid waste management process in semi-urban areas in India see [17, 24–27] lack of data and inconsistency in existing data is a major hurdle studying in developing nations like India. Semi urban area is very little information regarding solid waste produced in peri-urban areas unsatisfactory level of environmental services such as water supply the management of solid waste is going through a critical phase due to the unavailability of suitable facilities to treat and dispose large amount of municipal solid waste get generated daily metropolitan cities. Lack of financial resources, institutional weakness and improper technology and public apathy towards municipal solid waste are listed among the bottlenecks to provision of efficient and effective municipal solid waste management in India [24].

Municipal solid waste management has been found critical to public health and environmental improvements, urban areas of India became acutely aware of the problem in 1994, in the waste suspected in plague epidemic in Surat, an industrial city in the state of Gujarat. The first major attempt to develop a national strategy of solid waste management by National Environmental Engineering Research Institute [28], focused mainly the issues of urban areas with population more than 100,000. The Central Pollution Control Board study also reported widespread use of unnotified dumpsites for disposal of solid wastes in these towns. In spite of spending 30–50% of the total municipal budgets on solid waste management [29]. The unsatisfactory outcomes of current solid waste management services points to need for a sustainable solid waste management approach in semi-urban areas [27]. Delivery services is the another consequences of poor managed finance the failure of municipal bodies to deliver basic urban services. The management of solid waste in small towns in a particularly useful indicator of the efficiency of urban local bodies metropolitan cities are better provided with both water and solid waste management system then other urban and semi-urban centers [25]. The author also points out in Mirzapur (North-India small town) area, rickshaws piled high with waste can be seen careening through the streets, often through the streets, often depositing half of what they have collected on the road. The rest is thrown on the banks of the river Ganga that runs through the town. In Janjgir, even cycle rickshaws are not available. Men pulling handcarts clear the garbage. This naturally reduces the efficiency and frequency of collection. Many municipal bodies in small towns do not have the funds to transport solid waste to dumps outside the urban area. As a result, it is dumped within town limits. Hence, while in Mirzapur you see piles of garbage alongside the temples that dot the banks of the river Ganga, empty plots within town limits inevitably become garbage dumps in other towns.

2. Contingent valuation analysis

The Contingent Valuation Method (CVM) is a widely used non-market valuation method especially in the areas of environmental cost–benefit analysis and environmental impact assessment [30–32]. Contingent valuation is now used around the world in recent years, CVM has been extensively used in both developed and developing countries for valuation of a wide range of environmental goods and services (see [5, 33–37]). Ciriacy-Wantrup [38] had first proposed the contingent valuation method. Had discussed an individual should be interviewed and asked how much money they are willing to pay for successive additional quantities of collective extra-market good. If the individual values are aggregated the result corresponds to a market demand schedule (See [39]). Contingent valuation method of solid waste management research also emerged in developing countries Whittington et al. [37] Kathumadu in Nepal, [40] Gujranwala city in Punjab in Pakistan, Weldesilassie et al. [41] Addis Ababa Ethiopia, Murad et al. [42] and Chuen-Khee and Othman [1] in Malaysia, [43] in Yunnan Province China, Fonta et al. [44] in Nigera, Jianjun Jin et al. [45] Macao in China. India is very few studies to investigate the effect of waste of waste separation on the willingness to pay for improved waste management services for example Prasenjit Sarkhel and Sarmila Banerjee [46] adopted the contingent valuation method (CVM) with willingness to pay (WTP) of the household for waste management programme in a typical Indian Municipality the Ballay municipality in west Bengal including the willingness to pay questions, the contingent valuation questionnaire was divided into seven parts and the total number of samples were 570 and the mean willingness to pay from the responses to the open-ended questions was calculated 75% of the respondents expressing their willingness to pay at less than $ 1 per month regular waste collection in Bally the municipality in West Bengal. Sukanya Das et al. [47] had studied in the willingness to pay (WTP) for improvements in the solid waste management (SWM) services provided in Chandernagore and south Dum Dum municipality of Greater Kolkatta in West Bengal in this study 101 randomly selected residents took part in that choice experiment survey. Data were analyzed with conditional logit and random parameter logit with the interactions models. The study had revealed that on an average the residents of these municipalities were willingness to pay less than $ 1. While, this study had indicated that the public on average cared much about improvements in solid waste management in their locality.

3. Background of the study

Madurai has an area of 52 km^2 with in an urban area now extending over as much as 130 km^2 and it is located at show location on an interactive map 9°56'N 78° 07'E/ 9.93°N 78.12°E 19.93; 78.12. It has an average elevation of 101 m above mean sea level [48]. In Madurai city the daily generation of waste escalated from 360 tonnes in 2001 to 543 tonnes in 2011 [49]. The semi-urban waste generation per day 67 tonnes [50, 51]. **Table 1** shows that Avaniyapuram generates the highest waste generates among the major semi-urban areas in Madurai. Madurai city has a

Sl. no	Semi-urban areas	Male	Female	Total	Solid waste generation metric tonnes per day
1	Paravai	8346	8000	16,346	4
2	Vilangudi	10,640	10,433	21,073	2
3	Anaiyur	19,305	18,997	38,302	3.2
4	Avaniyapuram	27,099	25,907	53,006	17
5	Tiruparankundram	19,615	1939	39,009	14
6	Harveypatti	4089	44,046	8135	2
7	Thirunagar	7640	7909	15,549	1.3
8	A.Vallaiapatti	3529	3539	7068	2
9	Palamedu	4127	4060	8187	2.4
10	Vadipatti	10,875	10,905	21,780	4
11	Sholavandan	10,845	10,816	21,661	3.4
12	Alanganallur	5574	5490	11,064	3
13	Elumalai	7051	6979	14,030	2.8
14	Peraiyur	4512	4368	8880	2.5
15	T. Kallupatti	4857	4582	9439	3.4
	Total	**148,104**	**167,970**	**293,529**	**67**

Source: Author's calculation.

Table 1.
Semi-urban areas of Madurai District.

number of problems with collection and disposal of solid waste in general semi-urban areas in particular. First collection coverage is hugely inadequate, second lack of cost recovery and the unsustainable fee structure for current waste collection and disposal are serious issues. Solid waste management is one of the important obligatory functions of rural areas. However, this has not been efficiency performed by the urban local bodies Madurai rural solid waste generation has been significantly increase for example, generation of agricultural waste is 4.32 tonnes for every 3 months but did not properly reuse or recycling semi-urban and rural areas having more problems such as electricity, water supply, lack of ponds or through tube wells. What about solid waste lie uncollected along roadsides or if collected are dumped in an low-lying land. The practices are not only despoiling the local landscape but are an immense health hazard. The rapid growth of population in semi-urban areas in the last decade has meant that the volume of solid waste liquid waste has increases but the institutional capacities to handle them, remain absent [24, 25, 27]. This study has introduction of service charges for solid waste management has been received much attention among local bodies due to the continuous financial shortage of the local government for providing waste management services to an acceptable level. The pricing this service has expected to bring about efficiency as well as sustainability in providing this services.

Sampling and design of survey questionnaire.

4. Materials and methods

The study is confined to Madurai semi-urban areas. Madurai district (region) is existing 15 semi-urban areas (see **Table 1**). The sample units were selected adopting the stratified random sampling method. A total of 150 schedule 10 household respondents from each semi-urban area. The design of the survey followed recommendations from the NOAA panel on contingent valuation (see [52, 53]) and consist of two sections. Questions in the survey's first section asked about respondent's socio economic conditions in the household's survey section two questioned respondents about their willingness to pay. The hypothetical improved condition, and how each consumer would pay for the improved waste management services in Madurai (**Figure 1**).

The contingent valuation employed a single-bounded dichotomous choice format by open-ended questions in the WTP section. The survey was conducted March–April 2012. The survey was given to 150 randomly selected in Madurai semi-urban areas data covered socioeconomic characteristics of the household, including gender, age, marital status, education, household income, family size, employment and WTP for environmental improvement and better solid waste management. **Table 2** describes the variables.

4.1 Willingness to pay for improved waste management services in the study area

The **Logit regression** model had been used for studying about the probability of occurrence of an event by fitting a logit function. It is a generalized linear model used for binomial regression. The logit model was adopted since the Ordinary Least Square (OLS) producer was not appropriate particularly when the dependent variable is dichotomous. The problem with the OLS estimate however is the non- fulfillment of $O = (Y_i/X)$ since $E\ (Y_i/X)$ in the liner probability model measures the conditional probability of the event Y occurring given X_1 and must necessarily lie between 0 and 1 [54]. Like many other forms of regression analysis,

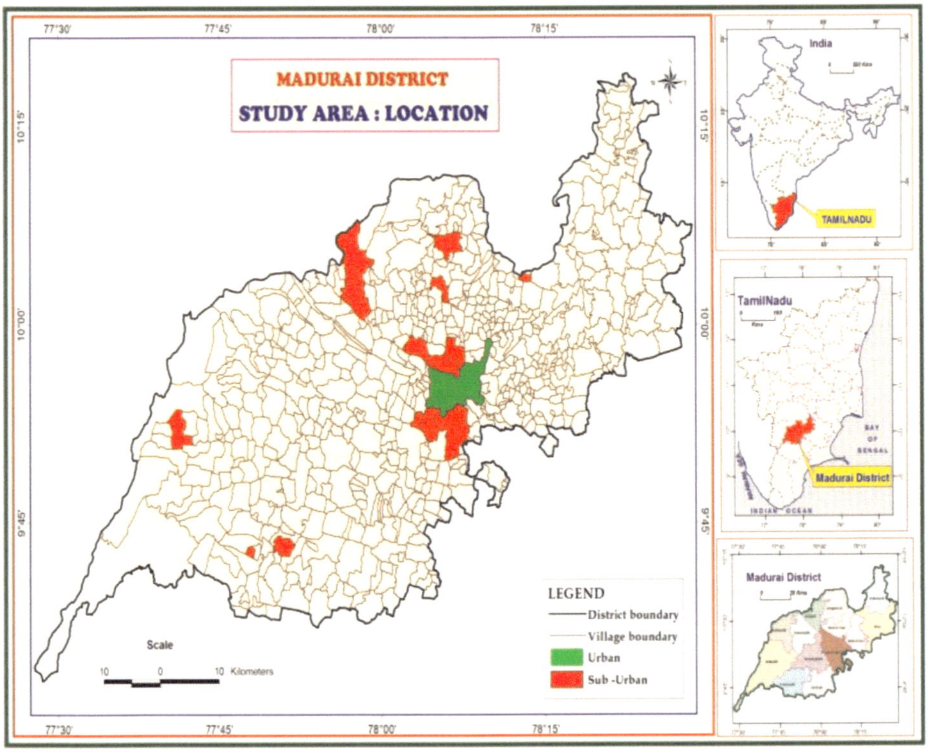

Figure 1.
Madurai semi-urban areas.

Variable definition	Description
Willingness to pay (WTP) de. var	1 if willingness to pay 0 otherwise
Age (AGR)	Age of respondent in year
Sex (SER)	Gender of respondent coded as 1 male for 0 female
Educational level (EDL)	Education of respondent represented as 1 for primary 2 for secondary 3 university level
Family size (Fam_Sz)	Number of members of the household
Income	Monthly income of the head of respondent in INR
Not satisfied	Are you satisfied for the present cleaning status if yes 1 and 2 no
Maximum amount willing to pay	Maximum amount of willing to pay for improved solid waste management

Table 2.
Description of the variables.

it makes use of several predictor variables that might be either numerical or categorical. This Study had applied the logit regression of willingness to pay for improved environmental quality, to determine the willingness of the respondents to bear the costs of improving the environmental quality in the study area. The Logit Model had been used to analyze the respondents' willingness to pay for an improved waste management service and the factors influencing their willingness to pay.

4.2 Willingness to pay for improved waste management services

To obtain the willingness to pay by the households for an improvement in their solid waste management, the responses of the households for willingness to pay was regressed on the socio economic characteristics. The coefficient estimates obtained for the WTP of the respondents (sex, age, education, family size, monthly size, monthly income, present cleaning status and maximum amount), the logit regression Model [55] was specified as

$$Y = \frac{1}{1+\exp^{-(\beta_0 + \beta_1 x)}}$$

where

Y = Response of households', sex, age, education, family size, monthly size, monthly income, present cleaning status and maximum amount of willing to pay for respondents to the willingness to pay question which was either

'1' if Yes or '0' if No.

β0 = is the intercept which is constant

β1 = is the coefficient of the price that the household are willing to pay for waste management services.

X = is a set of independent variable

4.3 Factors influencing willingness to pay for improved waste management services

To identify the factors influencing the willingness to pay of the sex, age, educational level, family size, monthly income, present cleaning status and maximum amount of willing to pay of the respondents for improved solid waste management, the respondents to the willingness to pay was regressed on the prices they were asked to pay and on the other socio- economic characteristics of the households. The logit regression Model was specified as

$$Y = \frac{1}{1+\exp^{-z}}$$

$Z = \beta_0 + \beta_1 x_1 + \beta_2 x_2 + \beta_3 x_3 + \beta_4 x_4 + \beta_5 x_5 + \beta_6 x_6 + \beta_7 x_7$

Y = Response of the education, occupation, household size, income, method of collection, agency of collection, amount they were willing to pay, for the respondents to the willingness to pay question which was either '1' if Yes or '0' if No.

$\mathbf{x_1}$ = sex (dummy: Male = 1, Female = 2),

$\mathbf{x_2}$ = Age (years),

$\mathbf{x_3}$ = Education (Dummy: Primary = 1, Secondary = 2, University level = 3),

$\mathbf{x_4}$ = Family size (numbers),

$\mathbf{x_5}$ = Monthly income INR (Indian rupees'00),

$\mathbf{x_6}$ = Present solid waste cleaning status if satisfied **(dummy: yes = 1, no = 0),**

$\mathbf{x_7}$ = maximum amount of willing to pay.

The pseudo-R square and the chi-square were used to measure the goodness of fit of the model and the significance of the model used.

Table 1 had depicted population and solid waste generation of semi-urban areas of Madurai district. Avaniyapuram has been highest waste generation in Madurai

	N	Minimum	Maximum	Mean	Std. devia
Willingness to pay	150	1.00	2.00	1.2267	42,008
Sex	150	1.00	2.00	1.1667	0.37393
Age	150	26.00	74.00	46.9600	11.82585
Education	150	1.00	3.00	1.2067	0.50894
Working status	150	1.00	2.00	1.2600	0.44010
Occupation	150	1.00	4.00	2.3000	1.07909
Monthly income	150	2300.00	18500.00	7865.2000	3902.80448
Family size	150	1.00	2.00	1.4400	0.49805
Maximum amount wtp	150	00	3.00	2.4067	0.72442
Present status	150	00	1.00	0.2933	0.45682

Source: Author's calculation.

Table 3.
Descriptive statistics.

compared to other semi-urban areas. This table clearly shows that the Thiruparankundram semi-urban areas are second highest population and waste generation. Semi- urban areas had been one of the more wastes contribution by these percentage are in Madurai district; per day total waste generation 67 metric tonnes as shown in table (**Table 3**).

4.4 Data description

Table 4 provides WTP responses in relation to the socio economic characteristics of the sample households. About 96% of the respondents had positive WTP values for the improvement in solid waste management services. The average monthly income of the sample households was INR 7865 with a minimum monthly income INR 2300 and a maximum of the INR 18500. The average of respondents was 46 years and average family size 1.44. Furthermore, about 50.7% of the respondents were willing to pay more than 100 for clean environmental services. While this survey had found that the highest percent of the respondents had primary education 84 and 11.3% respondents was secondary education level. Found this survey percent of the respondents mentioned the solid waste problems in their neighborhood to be one of the most urgent environmental problems.

5. Results and discussion

In this section, we present the discuss the result of the logit regression analysis to help determine which factors are significant for improved solid waste management services as well the amount respondents are willing to pay.

The 150 completed interviews, 4 respondents had invalid responses[1] to the valuation question. For only one variable was quite significant. CVM method suffers from one more problem, that is, how to estimate aggregate values based on the

[1] By invalid, an identified actual or protect or zero to the valuation questions by asking respondents not willing to pay for SWMS. In this respect 2 respondent had no faith in the scheme, 1 respondents had already paid some kinds of taxes to local government, and 1 had insufficient income

individual values expressed through willingness to pay. **Table 5** the logit results for the variables that are significantly related to the probability of providing positive WTP values. While sex, educational level, family size, present solid waste management system is not satisfied, and age, educational level and maximum willingness to pay are negative. The study had significant found that the sex is important significant factor for improved solid waste management services in the study area. The sex and willingness to pay services 10 percent level of significance.

This study had found that the age and willingness to pay no significant effect on the amount of willingness to pay for improved solid waste management services. WTP and educational is also no quite significant and the maximum amount of willingness to pay for improved solid waste management services are negative responses represent from respondents in the study area. Household income and willingness to pay for solid waste management services in important significant factor [2] this study had found that the income is insignificant for improved solid waste management services. As seen in **Table 4** low income people are interesting more willing to pay but it quite significant 74 respondent out of 150 are willing to pay for improved solid waste management services. The current solid waste management system is unsatisfactory in urban Indian in general and semi-urban in particular. This study had found that the 70.7 percent of the respondent are felt current solid waste management system is satisfied. Indian municipalities have overall responsibility for solid waste management their cities or local areas but most of the cities and semi-urban areas currently unable to fulfill their duty to ensure environmentally sound and sustainable ways of dealing with waste generation, collection, transport, treatment, and disposal. The failure of municipal solid waste

Socio economic variable		**WTP (Yes/No for improved solid waste management services)**	
		Yes	**No**
Gender	**Male**	**96**	**29**
	Female	**20**	**5**
Age of household head (in years)	26–36	29	7
	37–46	33	11
	47–56	28	11
	57–66	21	4
	Above 67	5	1
Education level	Primary	97	29
	Secondary	16	1
	University	3	4
Monthly Income	2300–8300	74	17
	8301–13,301	31	12
	13,302–18,302	8	5
	Above 18,303	3	0
Family size	2–4	62	22
	5–8	54	12
Employment	Government	34	12
	Private	30	8
	Self-employee	34	7
	others	18	7

Source: Author's calculation.

Table 4.
Willingness to pay person and socio economic characteristics of sample household.

Variables	Coefficient	Wald statistics
Sex (SER)	2.282*	6.595
Age (AGR)	−17.100	.000
Educational level (EDLR)	1.764	2.461
Family size (Fam_Sz)	.684	.991
Income	−21.495	.000
Not satisfied	.378	.554
Maximum amount willing to pay	−.265	.348

*Source: Author's calculation, * represent 10% level of significance.*
Log Likelihood 89.437, Number of observation 150.
Chi-square LR statistics 35.628, Significance 0.002.

Table 5.
Logit model estimation of willingness to pay for improved solid waste management services.

Willingness to pay		Amount of willing to pay		
Yes	No	<50	51–100	>100
96 (96%)	4 (4%)	3 (2%)	65 (43%)	76 (50.7%)

Source: Author's calculation.

Table 6.
Respondent willing to pay for solid waste management services.

management (MSWM) can result in serious health problems and environmental degradation.

Table 6 represents the respondent willing to pay for solid waste management services 96% of the respondents are willingness to pay but very less amount respectively only 2% of the respondents are INR Rs 50 ($1), 43% of the respondents are INR 100 ($2) and 50.7% of the respondents are more than INR 100 ($2) for improved solid waste management services.

6. Conclusion

This study finds that an average willing to pay about INR Rs. 24 (less than $ 1). This result of the study show that the demand for improved waste management is only significant related to the sex of the household respondent. Attempts must be made to improve willingness to pay solid waste management services in the semi-urban areas. To achieve this government should concentrate first on awareness campaigns about the consequences of waste mishandling and impacts of improper solid waste disposal. Previous studies have done only micro level analysis of economics of solid waste management did not improve any significant and scientific methodology adopted in solid waste management. Very few studies have been done in economics of solid waste management in India particularly contingent valuation analysis improved solid waste management services. Further, will need more empirical analysis in economics of solid waste management for better understanding about efficient solid waste management services are future good environmental services. Individual behavior and attitude as important environmental conservation and reduce solid waste generation. Future research in solid waste management should concentrate integrated with physiological factors of household recycling behavior and socio economic factors of solid waste generation.

Author details

Muniyandi Balasubramanian
Centre for Ecological Economics and Natural Resources, Institute for Social and Economic Change, Bangalore, India

*Address all correspondence to: balasubramanian@isec.ac.in

References

[1] Chuen-Khee P, Othman J. Household Demand for Solid Waste Disposal Options in Malaysia. World Academy of Science, Engineering and Technology. 2010;**66**:1153-1158

[2] Hagos D, Mekonnen A, Gebreegziabher Z. Households' Willingness to Pay for Improved Urban Waste Management in Mekelle City, Ethiopia. EfD DP Series, 12-06; 2012. Retrieved February 03, 2014, from www.rff.org/rff/documents/efd-dp-12-06.pdf

[3] Pearce D, Truner. Economics and Solid Waste Management in Developing World, the Centre for Social and Economic Research on the Global Environment (CSERGE). University college of London; 1994. pp. 1-21

[4] Foo TS. Recycling of domestic waste: Early experience in Singapore. Habitat International. 1997;**21**:277-289

[5] Wang H, He J, kim Y, Kamata T. Municipal Solid Waste Management in Small Town an Economic Analysis Conducted in Yunnan. China: Policy Research Working paper; 2011a. pp. 1-27

[6] Maria Eugenia Ibarraran V, Cortes II, Cuevas EM. Economic Valuation of the Environmental Impact of Solid Waste Management: A Case Study. University de as Americas-puebla;2001

[7] Medina M. Solid wastes, poverty and the environment in developing country cities. United Nations University, Working Paper Series No. 2010/23; 2010

[8] Zerbock O, Candidate MS. Urban Solid Waste Management Waste Productin in Developing Counties, School of Forest Resources & Environmental Sciences. Michigan Technological University; 2003. www.cee.mtu.edu/peacecorps

[9] Thomas-Hope E. Solid Waste Management: Critical Issues for Developing Countries. Kingston: Canoe Press; 1998

[10] Shekdar AV. Sustainable solid waste management: An integrated approach for Asian countries. Waste Management. 2009;**29**(4):1438-1448

[11] Beede DN, Bloom DE. The economics of municipal solid waste. The World Bank Observer. 1995;**10**(2): 113-150

[12] Wang H, He J, Kim Y, Kamata T. Municipal Solid Waste Management in Small Towns: An Economic Analysis Conducted in Yunnan, China. The World Bank; 2011

[13] UNEP (United Nations Environment Program). The Use of Economic Instruments in Environmental Policy: Opportunities and Challenges. Geneva: UNEP; 2004

[14] Idris A, Inane B Hassan MN. Overview of waste disposal and landfills dumps in Asian countries. Material cycles and waste management. 2004;**16**: 104-110

[15] Census 2011 Government of India

[16] Shree M. Keeping our cities clean: Urban solid waste management in Karnataka. Journal of Social and Economic Development. 2004;**6**(2): 159-175

[17] Balasubramanian M, Dhulasi Birundha V. Generation of solid waste in rural Tamil Nadu. Journal of Rural Development. 2011;**30**(1):101-112

[18] Jha MK, Sondhi OAK, Pansare M. Solid waste management case study. Indian Journal of Environmental Protection. 2003;**23**(10):1153-1160

[19] Kansal A. Solid waste Management for India. Indian Journal of Environmental Protection. 2002;**18**(2): 444-448

[20] Sunil K, Bhattacharya JK, Vahadilla Tapan Chakrabarthi AN, Devotta S, Akolkar AB. Assessment of the status of municipal solid waste management in metro cities, class I cities, and class II towns in India. Journal of Waste Management. 2009;**29**:883-895

[21] Mahmood SKAG, Trivedi RC. Municipal solid waste management in Indian cities – A review. Journal of Waste Management. 2008;**28**: 459-467

[22] Mahmood SMAK. Analysis of municipal solid waste management systems in Delhi – A review. In: Book of Proceedings for the Second International Congress of Chemistry and Environment, Indore, India. 2005. pp. 773-777

[23] Ray MR, Roychoudhury S, Roy MG, Lahiri S. Respiratory and general health impairments of workers employed in a municipal solid waste disposal at open landfill site in Delhi. International Journal of Hygiene and Environmental Health. 2005;**108**(4):225-262

[24] Show A. Peri-urban interface of Indian cities growth, Governance and Local Initiatives. Economic and Political Weekly. 2005;**40**(2):129-136

[25] Sharma K. Rejuvenating India's small towns. Economic and Political Weekly. 2012;**XLVII**(30):63-68

[26] Srinivasan K. Public, private and voluntary agencies in solid waste management: A study in Chennai city. Economic and Political Weekly. 2006: 2259-2267

[27] Sundaravadevel M, Vignewaran s, Doeleman JA. Waste management in semi-urban areas of India: Appropriate technological strategies to overcome financial barriers. Environmental Engineering and Policy. 2000;**2**:91-104

[28] NEERI. Strategy Paper on Solid Waste Management in India. Nagpur: National Environmental Engineering Research Institute; 1995

[29] Pannerselvam L. Community Based SWM Project Preparation. Proceedings of the 21th WEDC Conference, Columbo; 1994

[30] Cummings RG, Brookshire DS, Schulze WD. Valuing Environmental Goods: A State of the Arts Assessment of the Contingent Valuation Method. Totowa, NJ: Roweman and Allanheld; 1986

[31] Mitchell RC, Carson RT, Ruud PA. Cincinnati Visibiliry Valuation Study: Pilot Study Findings. 1989. Report to the Electric Power Research Institute

[32] Venkatachalam L. The contingent valuation method: A review. Environmental Impact Assessment Review. 2004;**24**:89-124

[33] Carson RT, Flores N, Meade N. Contingent valuation: Controversies and evidence. Environmental and Resource Economics. 2001;**19**(2):173-210

[34] Carson RT. Contingent Valuation: A Comprehensive Bibliography and History. Cheltenham, UK: Edward Edgar; 2002

[35] Carson R, Flores NE, Hanemann WM. Sequencing and valuing public goods. Journal of Environmental Economics and Management. 1998; **36**(3):314-323

[36] Tait PR, Friesen L, Cullen R. Will unit pricing reduce domestic waste? Lessons from a contingent valuation study. New Zealand Economic Papers. 2005;**39**(1):83-103

[37] Whittington D, Briscoe J, Mu X, Barron W. Estimating the willingness to pay for water services in developing countries: A case study of the use of contingent valuation surveys in southern Haiti. Economic Development and Cultural Change. 1990;**38**(2): 293-312

[38] Ciriacy-Wantrup SV. Capital returns from soil conservation practices. Journal of Farm Economics. 1947;**29**: 1181-1196

[39] Hanemann MW. Valuing the environment through contingent valuation method. Journal of Economic Perspectives. 1994;**8**(4):19-43

[40] Altaf MA, Deshazo JR. Household demand for improved solid waste management: A case study of Gujranwala. Pakistan. World Development. 1996;**24**(5):857-868

[41] Weldesilassie AB, Frör O, Boelee E, Dabbert S. The economic value of improved wastewater irrigation: A contingent valuation study in Addis Ababa Ethiopia. Journal of Agricultural and Resource Economics. 2009;**34**(3): 428-449

[42] Murad MW, Raqub MA, Siwar C. Willingness of the poor to pay for improved access to solid waste collection and disposal services. The Journal of 68 Environment and Development. 2007;**16**:84. SAGE Publication. http//www. Sagepublication.com

[43] Wang H, He J, Kim Y, Kamata T. Municipal Solid Waste Management in Small Towns: An Economic Analysis Conducted in Yunnan, China. Policy Research Working Paper 5767. Washington, DC: World Bank; 2011b

[44] Fonta MW, Ichoku HE. The application of the contingent valuation method to community-led financing schemes: Evidence from rural cameroon. Journal of Developing Areas. 2005;**39**(1):109-126

[45] Jin J, Wang Z, Ran S. Comparison of contingent valuation and choice experiment in solid waste management programs in Macao. Ecological Economics. 2006;**57**(2006):430-441

[46] Sarkhel P, Banerjee S. Municipal solid waste management, Source – Separated waste and stakeholder's attitude: A contingent Valuation Study. Environment, Development and Sustainability. 2009;**12**(5):611-630

[47] Das S, Birol E, Battacharya RN. Informing efficient and effective solid waste management to improve local environmental quality and public health: Application of the choice experiment method in West Bengal, India. Discussion Paper, University of Cambridge. 2008;**33**:1-25

[48] Alaguraj, Durairaju S, Yuvaraj D, Sekar M, Muthuveerran P, Manivel M, et al. Land use and land cover mapping – Madurai District, Tamilnadu, India using remote sensing and GIS techniques. International Journal of Civil and Structural Engineering. 2010; **1**(1):91-100

[49] Annopu RK. Sustainable Solid Waste Management in India, MSc Dissertation. Columbia University in the City of New York; 2012

[50] Balasubramanian. An economic study of solid waste management in Tamil Nadu with special reference to semi-urban areas of Madurai. Unpublished Ph.D Thesis; 2011a

[51] Balasubramanian. Municipal solid waste management current problems and elucidation in India. Productive Journal. 2011b;**51**(4):308-314

[52] Arrow JK, Solow R, Portney PR, Leamer EE, Radner R, Schumand H. Report of the NOAA panel on

contingent valuation. Federal Register. 1993;**58**(10):4601-4614

[53] Mitchell RC, Carson RT. Using Surveys to Value Public Goods: The Contingent Valuation Method. Washington, DC: Resources for the Future; 1989

[54] Gujarati DN. Basic Econometrics. 2nd. ed. New York: Tata McGraw-Hill Book Company Inc; 1988. pp. 467-490

[55] Gujarati DN. Introduction to Econometrics. New York: McGraw-Hill, B.E.F.E.; 2006

Section 4

Disposal of Municipal Solid Wastes

Chapter 10

Municipal Solid Waste Disposal in Mangrove Forest: Environmental Implication and Management Strategies in the Niger Delta, Nigeria

Aroloye O. Numbere

Abstract

Niger Delta is an oil rich region situated in the southern part of Nigeria. It is made up of nine states which hosts oil industries. There are a handful of businesses (super market, manufacturing companies, etc.) that service the over 40 million people living in the cities. This situation had led to the increase in solid waste in the city. Because of the problem of over population, and poor waste management strategies (e.g., lack of recycling habit and lack of equipment) the mangrove forest had become a dumping ground for waste. This action has impacted the health of aquatic and terrestrial organisms, and has created a public health disaster for citizens because of increase in heavy metal concentration up the food chain. This chapter therefore, identifies poverty, lack of planning, poor behavior and poor technology as key factors affecting effective waste management in the Niger Delta. It suggests that good waste management system can be worked out if there is coordination between research institution and government in the implementation of recommendation by research institutes. Attitudinal change is also necessary on the part of citizens and government to enable a healthy interaction for the purpose of managing waste effectively.

Keywords: solid waste, mangrove forest, recycling, heavy metal, open dump, Niger Delta, city planning

1. Introduction

Municipal refers to a city, but when a city is very large it becomes a metropolis. According to [1] government performs two kinds of functions: (1) supply of goods and services within municipality and (2) conflict management. For instance, the sitting of garbage dump site can lead to conflict, which is to be resolved by the municipality. Other functions of a municipal government are police, fire protection, and street maintenance. In the area of conflict management municipal government takes care of the sitting of garbage dump sites within the city to prevent clash of residents over land ownership rights. The government also decides on the number of garbage collection per week, and the disposal mechanisms to ensure a clean and healthy city for people to live in.

Waste on the other hand, is an unwanted residue that is no longer useful to the system, but useful to another system. Municipal solid wastes are non-liquid wastes that are by-products of manufacturing and processing industries that are within the city. Solid waste are every day materials and occur as follows: paper (50.7%), food waste (19.1%), metal (10.0%), glass (9.7%), wood (2.9%), textiles (2.6%), leather and rubber (1.9%) plastics (1.4%) and miscellaneous (1.7%). Lack of monitoring of waste movement in municipality can lead to indiscriminate disposal of waste, for example, waste is dumped without restriction in mangrove forest and rivers **Figure 1**. This is a problem that has been noticed in the course of our field work, but had remained unreported in the literature. This work is thus one of the first to report the problem of refuse disposal in mangrove forest in the Niger Delta. Mangroves are resilient [2, 3] and could withstand some level of pollution [4]. They are also a zone of high litter decomposition [5] as a result of the proliferation of microbial activities on forest floor [6]. This ability had made the mangroves to survive in the face of intense environmental pollution [4], but the effect of waste disposal on mangrove growth remains to be seen. This is because harmful heavy metals from non-biodegradable substances in the waste (e.g., plastics) can impede their growth. However, mangroves act as natural environmental biogeochemical barriers to pollutants generated in solid wastes disposal sites through mechanisms occurring at root level [7]. Mangrove roots produce oxygen to cope with the anaerobic condition of the soil. The creation of oxidized rhizosphere fixes heavy metal under non-available forms [8]. The large adventitious root system also restricts the movement and physical distribution of heavy metals. This prevents pollutant remobilization. Mangrove sediments effectively retain heavy metals by preventing migration. The heavy metals are prevented in the rhizosphere sediments under very refractory chemical forms, unable to be absorbed by plant roots. This thus, blocks the intoxication of the mangrove trees [7]. In addition mangroves root have a shutting down mechanism, which prevents the absorption and uptake of harmful pollutants, just as it shuts down the intake of excess salt when in saline environment.

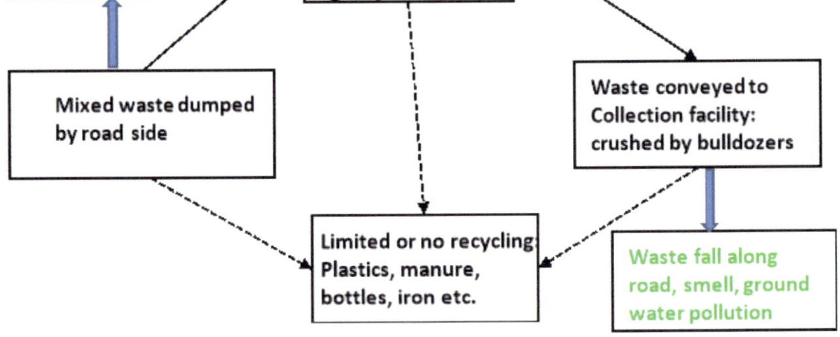

Figure 1.
Waste management strategy in cities in the Niger Delta, Nigeria. It shows the inefficiency in waste management leading to individualistic management resulting to mangrove forest and river pollution.

The disposal of solid waste in mangrove forests and wetland areas is as a result of ignorance of its health effect. It is also due to overwhelming production of waste from highly populated city dwellers with little or no technology to handle the waste surge. High generation of waste is usually prevalent amongst the low income class people who generate more waste than they can handle, they thus resort to self-help by disposing the waste themselves into drains during rain fall, into river and in mangrove forest. This situation had been going on for many years, especially in areas of the city such as waterfront and coastal towns inhabited by people of low income class. Those living at the water front are not considered in municipal waste disposal planning. They are also not considered in the planning of waste collection within and around the city. Since they are left out, they manage their own waste by themselves. This is because the waste management design in the region is faulty, and is the cause of negative feedback of excess unmanageable waste, which litters the streets of the city (**Figure 1**). This chapter has identified some problems of effective waste management, and had proffered some solutions towards resolving them.

2. Causes of high waste generation in cities in the Niger Delta

There are several causes of high waste generation in municipal areas in the Niger Delta, some include: (1) over population (2) poor town planning (3) lack of technology (4) Poor waste management habit and (5) Lack of sorting and recycling culture. Details of these factors are given below.

2.1 Over population

Nigeria's population is over 200 million making it the most populous black nation on earth. Out of this number, one-fifth (i.e., 40 million) lives in the southern part of the country known as the Niger Delta region [9]. Each family in the region is made up of an average of six persons. The waste production per person, multiplied by the entire population gives a staggering figure of waste. In the Niger Delta, one person produces approximately 1 kg of waste per day, multiplying this figure with the population size gives an approximate waste load of 40 million kg produced each day for the entire region. With inadequate facility, lack of manpower and poor waste management strategy, it is only one fifth (8 million kg) of the waste that is eventually evacuated daily leaving behind 32 million kg of waste that is not collected and left to litter the streets of the city. Abandoned wastes are scattered by rain and wind and carried into public drains. This blocks the free flow of water leading to flooding.

Another cause of the increase in population is the migration of people into the cities of the region in search of white collar jobs, which had led to the multiplication of the number of persons already in the city. The Niger Delta serves as the treasure base and the melting pot of the nation because of the abundance of industries, which attracts people from all works of life. It has refineries, oil companies, ports authority, fertilizer company, cement factory, etc. This had attracted people from other regions to migrate into the city in search for jobs. The average rural-urban migration rate is about a 1000 persons per day, which further increase the population. The numbers of persons that come into the cities outnumber the number of persons that go out of the cities. Out of the population that comes in about 2% remain and look for jobs or start small scale business, which generate waste. This adds to the total waste load of the cities in the Niger Delta. Furthermore, lack of data of the population growth rate, migration and emigration rate and waste production had complicated the process of effective waste management by waste

Figure 2.
(a) Waste dump site near former mangrove forest taken over by nypa palms at Eagle Island, Niger Delta, Nigeria. (b) Plastic waste in early recruiting mangrove seedling (Rhizophora spp.) brought in by tidal current. (c) Waste washed ashore by tidal current at Eagle Island. (d) Plastic waste recovered from the sea near Eagle Island placed near a mangrove forest awaiting transfer to plastic recycling company.

management agencies in the region. This is because for effective waste management each region should have been divided into sub-districts, and the population known to enable waste managers to understand the waste dynamics. This will help them to plan the number of trips of waste trucks that can completely evacuate waste from the district. Improper waste management in the face of over population is the main reason why people have resorted to self-help, by disposing waste in swamps, rivers and mangrove forest (**Figure 2a**).

2.2 Lack of proper town planning practice

Planning simply means to clarify one's objective and to determine actions that should be taken by whom, when, by what methods and at what costs in other to achieve the desired goal. It is also the evaluation of alternative choice, strategy, solution, plan, implementation and review. Planning is functional when it develops an appropriate course of action for decision makers. The purpose of planning is to provide for the "urban citizen" an environment suitable for human habitation. This is because the price to pay for lack of planning is huge and can affect the waste management process [10]. Lack of municipal planning can lead to the proliferation of slums, congestion, noise, waste, air and water pollution, overcrowding, inadequate school, unemployment, inadequate municipal services (e.g., waste disposal sites and recycling facilities), disease proliferation, crime, ugliness and a host of other societal ills.

Adequate physical planning aims to control physical development of communities thus, avoiding the major social and economic cost of non-planning such as inadequate waste management system leading to unhygienic conditions and flooding in cities, which erodes major road network. Furthermore, physical planning bears a distinct relationship to many other governmental functions [10, 11].

For instance, the decision about the patterns and locations of waste processing facility will involve not only the local planning agency, but also federal and state ministry of environment, environmental protection agency, public works and city engineers. This is because the location of a waste facility will have ripple effect on other facilities in the city, for example it will have an impact on the location of other land uses, and will affect the land use policies of communities throughout a given metropolis [12, 13]. Decision about the proper location of industry can have a major impact on employment and income levels, and thus the buying power and waste generation capacity. All of these interrelationships imply a need for some coordination amongst different municipal activities. Planning theory postulates standard for the location and space requirement of different land uses such as waste facility, housing, esthetics, recreation, industry, etc. on "design concept" [14]. There is a belief that different land uses should be kept separate, and density low. It is believed that an improvement in the physical will lead to an improvement in all the social and economic problems besetting urban communities.

A classic example of the nature of planning is the "zoning ordinance", which is a municipal law that divides the municipal area into district, within the district standards and restrictions are established for the use of land. For instance areas are designated as residential, industrial, recreation, and municipal waste collection facility (MWCF).

Waste generation can also be affected by pattern of street layouts. This is under the purview of 'sub-division regulation", which is a municipal law that controls the development of new residential area. This involves the width and pattern of streets, size of drainage facilities, sewage and water system and waste disposal site. Building code can also be used to manage waste generation and disposal system. It is an implementation device of planning. The enforcement of code is providing for the municipal standards for the structures and facilities for building as part of the municipal. The purpose of code enforcement is to safeguard health. Modern method of city planning is the use of Geographical Information System (GIS) to identify locations for establishing waste disposal facilities that will not affect esthetics and property value [15].

2.3 Lack of municipal waste recycling and treatment facilities

Central waste treatment and recycling facility is important in managing waste coming from different parts of the city [16]. The problem is that in the Niger Delta this facility is non-existent, thus waste collected from several locations are usually disposed off on open virgin land, crushed with bulldozers and mixed with soil to form compost (**Figure 1**). The waste is picked up by the side of the road and driven to such locations. Since the waste generation data is not available it becomes a problem for those evacuating the waste to know the carrying capacity of the trucks and the number of trips to go. This leads to the overloading of the pay load resulting in the waste falling off on the road when the waste is being driven to the crushing facility. In addition, after collection some waste still remain behind on the road for days because the truck had been filled beyond capacity. This type of waste management practice is open disposal, which is ancient and had been phased out in many parts of the world. This method is unhygienic because it is often situated around human habitation. It introduces pest and diseases through rodents and flies. The smell coming from such location is nauseating. This reduces the esthetic values of the city. The idea of establishing recycling facilities across the Niger Delta had been in the drawing board for decades without being implemented, which is as a result of bureaucratic bottlenecks in government. Currently it is only private investors that are making attempts to establish such facilities. Plastic products are the major recycled waste product in the region [17] (**Figure 2d**). This is because of the millions of plastic materials that are evacuated from the surrounding drainages and water bodies (**Figure 2b, c**). The reason for high plastic waste retrieval from the environment is because of the monetary reward of $0.3 offered for I kg of plastic waste recovered. In fact, the most priced plastic materials in the region are those that are made up of high density polyethylene (HDPE), which is used in the production of plastics, bottles, corrosion-resistant piping, geo-membranes and plastic lumber, etc. The retrieved materials are compressed and exported out of the country to generate foreign exchange.

2.4 Poor waste management habit

There is no waste sorting culture in the region, which creates problem in the effective management of waste. This is because collected waste at different sites are all mixed up at home by residents and dumped by the road side for collectors to pick. There is no separation of waste into different types as practiced in developed societies (**Figure 3a, c**). Lack of waste separation creates problem for collectors who do not have the training or the equipment for separating the waste into its parts before final disposal in landfills [18]. Gross ignorance and helplessness in waste management had made a lot of people to become waste distributors, who pass on waste from one place to another in the name of management. They do this be pouring waste in restricted places under the cover of darkness or sometimes in the open without being confronted. Favorite areas of waste disposal in some cities in the

Figure 3.
(a) Giant silo bin used to collect and convey building waste at a building renovation site. (b) Indoor waste disposal unit installed in a room to collect house hold waste, which will be channeled out into a silo bin placed outside for onward collection and disposal by waste agent at a building facility in Saint Louis USA. (c) Type I (rubbish, e.g., paper) waste collected and sorted for recycling (green silo).

Niger Delta are along roads. This is done to seek attention from the government or waste agency (**Figure 2a**) to come and evacuate pile of waste in their neighborhood. They also dump waste during heavy rainfall inside drains so that it will be flushed away by water into adjoining river. Wastes are also poured on farm and mangrove forest to conceal their action from the municipal authority. A favorite place where waste is usually dumped is at the foot of plantain and banana trees. This is because there is an erroneous belief that the waste act as manure for the growth of the plants. Dumping of waste in drainages blocks the free movement of waster leading to flooding problems. People also buy goods they do not need, which end up at the dump site. Poor income makes some people to buy sub-standard products that have low life span, resulting to wastage. These products become non-usable when used for a short time leading to their disposal in refuse dump.

2.5 Poor recycling culture

There are a lot of people who do not understand the principle of recycling, and do not care to know its importance. Rather local people are more interested on how to survive the difficult times. Some people think recycling is to pile up unused goods in their store house rather than giving it out to a recycling agency for the manufacture of new goods. Recycling is the re-use of waste for other beneficial products (**Figure 3c, d**). Recycling is beneficial in several ways: (i) it provides jobs, (ii) reduces waste volume for disposal, (iii) extends the life span of a land fill and (iv) used to manufacture new products. People do not have the habit of recycling goods because of their belief in conservatism in the ownership and use of goods. This means they buy only what they need, which help to prevent wastage of resources. They therefore buy goods they can consume without much left over. The method of gathering and disposing waste also makes it extremely difficult to recycle the waste. It is very difficult to sort and re-use a combination of liquid and dry waste in a dump site. Most refuse dump sites are made up of all waste types, i.e., type 0–type 6 wastes, which include a combination of paper, broken bottles, metal, wood, food items, hospital waste, kitchen waste, etc. The mixing of the waste at the beginning and the combined disposal of the waste had made the sorting process to be very difficult. This overwhelms the waste agents who have no option than to process the waste as it is, using bulldozers rather than carrying out a recovery operation for the purpose of recycling [19]. In the Niger Delta, the major recycled waste is plastics (**Figure 2d**). This is common because of the monetary value attached for its recovery, and a major driving force is poverty. This is because people that are well to do not scavenge for plastics products for financial gain. Rather many jobless individuals had made it their job by searching for plastics in every conceivable place such as drainage, river, swamp and refuse dump sites. They take great risk to collect plastic products and send them to the manufacturing companies who use them for producing plastic products for pecuniary gains. These groups of scavengers are sometimes destitute who have no home but sleep on the streets. They take great risk to their health and lives to recover the plastics and sometimes bottle products by using their bare hands or iron rod to rummage through the piles of refuse at the dump sites.

2.6 Lack of technology

The problem of third world nations is technology, which affects the pace of development in all fields of life, and waste management is not an exception. The ancient method of waste disposal being open dump is still practiced in many places in the region. In this methods open trucks are still used to evacuate waste from the streets. The disadvantage of this method is that it leads to the scattering of waste materials along the streets of the city. This occurs when open trucks are used to convey waste to waste collection facility. The trucks are often overloaded leading to some of the waste being blown away by wind thereby littering the streets. It is also unhygienic for motorists who are made to endure the stench when the truck comes closer to them. The modern method of using home garbage receptacle such as trash chutes, (**Figure 3b**) silo bins (**Figure 3a, c**) and silo disposal trucks is not used due to paucity of funds to acquire them for use in the municipality. This method was used in the past, but because of lack of maintenance culture and continuity in governance has made the whole equipment that was initially acquired by previous governments amounting to millions of dollars to go moribund after its abandonment. Presently there are constructions of concrete waste disposal sites around the cities where people go to dump their waste. This method is also problematic because it is still an open dump, which reduces the esthetic value of the city. Waste materials can also be carried away by wind and water especially during heavy rainfall leading to flooding. It discourages waste sorting because the different wastes are mixed up before their deposition at the dump site (**Figure 2a**).

3. Waste management strategies in municipal areas in the Niger Delta

Different cities in the Niger delta have their way of managing waste, but collectively the major management methods adopted in most areas is individual and group management. Individual residents manage their garbage and trash. Different occupants of a house work as a group to manage their waste system. For instance, they perform rotational sweeping, collection, dumping and burning of waste in open spaces designated for that purpose. People living in a given area contribute money, which they use to hire and pay agents or private refuse collectors. The waste are put in drums or bins and later disposed off by the paid agents, who comes weekly to collect and dump the waste in approved dump site. Local government manages waste in their various jurisdictions especially in their headquarters. The local government has the constitutional role of waste management, which is part of their social responsibility. Market operators and indigenous manufacturers also manage their own waste. The problem is that private individuals use open spaces such as mangrove forest or wilderness (**Figure 2a–c**) as sites to dispose off their waste. This is because they do not have waste collection and disposal system. They dump waste such as animal carcasses, metal scrap, vehicle junks, plastics, etc. Similarly, domestic, industrial and biomedical wastes are all dumped together at dump sites. Hazardous and radioactive wastes are often dumped together, which is dangerous to public health and safety. Domestic and industrial liquid wastes are indiscriminately discharged or find their way into streams and rivers, which serve as drinking water for a large section of the rural and urban inhabitants.

4. Reasons for ineffective waste management practice in Nigeria

There are several reasons for poor waste management in Nigeria, they include, lack of reliable research data on waste. There were limited data in the past, but of recent more scientist and waste management scholars had written their dissertations on waste management. However, the problem is that the results of these studies had not been adopted and implemented by the government; the researches only end up in the shelves of libraries in the various high institutions. For proper waste management, in Nigeria there should be a meeting point between theory and practice. Agencies should be set up to create a cross fertilization of ideas between higher institutions and government agencies. This will help in the implementation of recommendations from the different studies. This is because many of the studies had gone to great lengths to collect long term data, which if implemented will help government in planning for effective waste management system for the municipalities and entire region at large.

Lack of sorting culture is also impeding the progress of waste management in the Niger Delta. People do not take it as a responsibility to separate their waste before disposal. They feel that it is a waste of time since the receivers of the waste do not care and will not sort the waste at their collection facility. Irregular collection and unhygienic disposal by private waste collectors are also some problems of ineffective waste management. Poverty is an overriding factor that had affected waste management. This is because even when rules of proper waste disposal are stipulated by the government many people do not follow such instructions because they lack the money to buy the waste containers for the collection of their waste. Government alone cannot be held responsible for poor waste management because many people exhibit poor attitude towards waste collection by not wanting to pay for waste management services. This may be attributed to lack of awareness on the dangers of improper waste management, and lack of community awareness of the economic value of waste recycling. There is also a craze for fashion which has made people to generate more waste than they can manage; especially women who adopt some fashion trend that is antagonistic to the local culture in terms of clothing and beautifying materials, which they later dispose into open dump or drain. A typical example is artificial hair, which in recent times had littered the streets of most cities. Similarly, make-up chemicals are flushed down the drain and can enter the river thus polluting the surface and ground water systems. This has a negative feedback because it can come back to humans through the food chain or drinking water causing health effects. Lastly, apart from the inaction of government and the poor attitude of citizens, there is also nonchalant attitude by industrialists and manufacturers, who are more interested in making profits than giving back to their host communities through the provision of social amenities such as waste bins and payment for waste evacuation.

5. Mangrove forest as refuse dump site

Waste get into mangrove forests through two means: (1) through tidal flushing (**Figure 2b, c**) and (2) through disposal by humans (**Figure 2a**). Tide washes ashore buoyant debris from far and wide. The debris accumulates at the edges of the sea and inside mangrove forest. These kind of waste are usually materials picked by tides from elsewhere such as leaves, branches of trees, plastics, carcasses, etc. (**Figure 2c**) while waste dumped by humans are mainly municipal waste such as household and industrial items such as food, paper, clothes and plastics, industrial waste, agricultural waste and market or commercial waste, majority of which are made of organic products [20].

Mangrove forests are found at the interface between the land and the sea. They are thus recipients of waste from both the land and the sea. They are usually seen and considered as waste land because of activities that go on in the mangrove forest are often not supervised. People who dispose refuse or cut the trees are not punished making others to do the same. Mangrove forests serve as homes for many people who clear and erect their houses. Those who live close to or inside the mangrove forest dispose their waste right in the forest since waste management agents do not come to evacuate their waste. Since mangrove areas are not under the jurisdiction of waste management agents, the people manage their waste by throwing them wherever they want. Mangrove trees are also cut and used as firewood to generate energy [21] and their cuttings act as waste that litter the forest floors.

6. Comparison of heavy metal concentration in soils at dump sites situated in mangrove and non-mangrove forest

6.1 Materials and methods

A study was done to compare the heavy metal content of soils collected from dump sites located in mangrove and non-mangrove forests. A total of 16 soil samples were collected randomly in a block design. The soil samples were collected 5 cm below the surface with a soil augur and placed in a cellophane bag and transported in cooler to the laboratory for physicochemical analysis.

6.2 Results

The result indicates that there was no significant difference in heavy metal concentration ($F_{1, 58} = 0.24$, $P > 0.05$) in soils collected from dump sites in mangrove and non-mangrove soils. Copper, lead and zinc had higher concentrations in non-mangrove soils whereas cadmium had the highest concentration in mangrove soil (**Figure 4**). Dump sites contain all kinds of waste such as domestic, industrial, hospital, municipal and electronic waste (e-waste). These materials had made the heavy metal concentration to be high in both soils, which is not good for the ground water aquifer and organisms that inhabit the soil. Arsenic is a poisonous chemical that has teratogenic effect on man. The result of heavy metal concentration from e-waste in mangroves and farm soil further illustrate the ability of mangrove forest to retain high heavy metal concentration, which is far above the required standards (**Table 1**). Waste materials that have high arsenic content can be disastrous to organisms that reside in the forest. The problem is

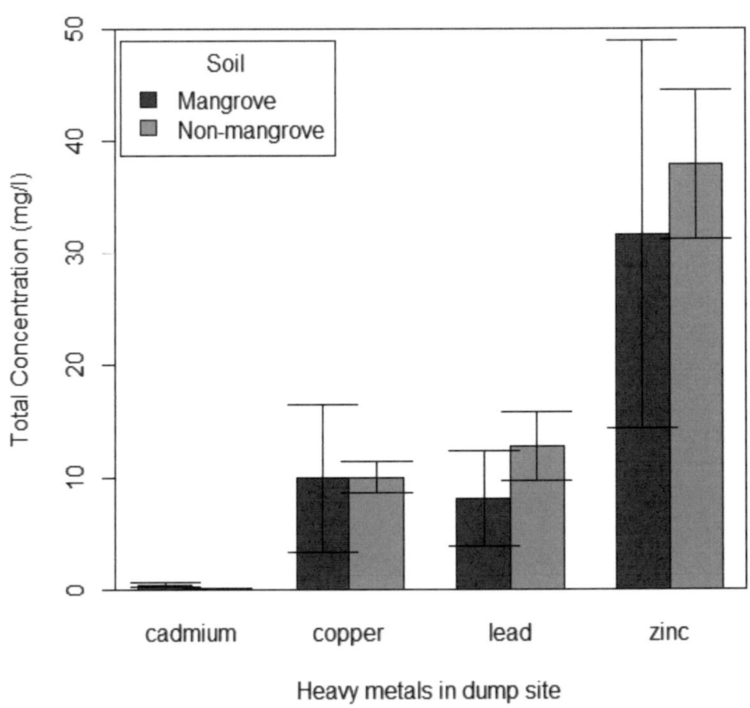

Figure 4.
Heavy metal concentration in soils collected from dump sites situated in mangrove and non-mangrove forest in selected sites in the Niger Delta, Nigeria.

Heavy metals	FMENV limit (mg/l)	Farm soil	Mangrove soil
		Conc. (mg/l)	Conc. (mg/l)
Arsenic	0.05	2.31 ± 1.41	0.80 ± 0.47
Cadmium	0.01	0.04 ± 0.03	0.09 ± 0.04
Lead	0.05	3.38 ± 0.74	4.36 ± 0.88
Zinc	0.05	23.62 ± 2.85	21.32 ± 6.75
Copper	0.05	16.77 ± 11.48	21.86 ± 18.20
Nickel	0.05	3.55 ± 1.06	23.18 ± 14.47

Table 1.
Comparison of heavy metal concentration from study and Federal Ministry of Environment (FMENV) maximum concentration for ground water protection.

that some organisms that are found in the mangrove forest are consumed by man. For example crabs, periwinkle, fish, etc. Another study using crab shell and tissue (*Goniopsis pelii*) show that the distribution of heavy metals in the body parts was highest in claw tissue: zinc (1894.5 mg/l), cadmium (28.0 mg/l), lead (283.0 mg/l) followed by gills: zinc (116.0 mg/l), cadmium (12.0 mg/l), lead (173.5 mg/l), and gut: zinc (38.0 mg/l), cadmium (2.8 mg/l), lead (27.4 mg/l). This can be attributed to accumulation of heavy metals that come from e-waste (mobile phones) in mangrove forest soil (*in press*). This can lead to biomagnification in man thereby causing health problems. It is also environmentally damaging when pollutants enter the food chain.

7. Discussion

Management of solid waste is a problem for most cities in Africa. This chapter has discussed four key causes of solid waste management problem in Nigeria, they are: behavioral, poor technology, poverty and poor town planning amongst others. The lack of knowledge in the management of waste by individuals magnify at municipal level leading to larger waste management problems. The best strategy to tackle this problem is a change of attitude of individuals, through deliberate decision to do things right. In addition, government can assist by embarking on intensive enlightenment campaign and provision of sophisticated waste evacuation equipment. Furthermore, stiff penalties should be put in place for defaulters as a means of deterrence (**Figure 5**). This is because poor technology and lack of technical skill are problems of third world countries. This makes it cumbersome to manage waste effectively in a highly populated country such as Nigeria. Nigeria does not lack manpower, but lacks the technology to adequately manage solid waste. Presently attempts are being made by private individuals to establish waste management and recycling facilities in several parts of the country. Similarly few state governments are making attempts at establishing waste treatment facilities in their states. It takes the will power and availability of funds to accomplish this aim.

Poverty is a problem that affects almost every aspect of life in the country, even when there is the will to pursue good waste management methods, the funds to acquire equipment to carry out the process is lacking. The establishment of a waste treatment facility is a gigantic project that requires government assistance and input to be successful. This is because of other aspects of the project that require huge financial commitment, e.g., purchase of land, good road network,

Figure 5.
Enlightenment campaign against waste disposal in rivers and a mangrove forest at Eagle Island, Niger Delta, Nigeria.

proper city planning and employment of skilled waste managers, which go beyond what an individual can do. However, the government is well positioned to execute such gigantic projects. For government to acquire land and to use the state's resources to establish waste facility will not be a problem as compared to what an individual will do. For example, because of the land use system, land acquisition by a private person is more cumbersome and costly than when done by the state government.

Planning also plays a key role in proper waste management. Most communities in Africa are built from communal land allocation without proper design such as surveying and land allocation to ensure the sitting of important municipal facilities at specific location. This makes the distribution of properties and facilities to be haphazard. For instance there are some places that are difficult to locate because there are no street names or street numbers. This affects the sitting of waste facility because central areas that are supposed to be preserved for public use have been taken over and houses built by private individuals. Lack of town planning thus affects the sitting of projects that will be beneficial to the people. Furthermore, solid waste treatment facilities are usually sited in locations that are inaccessible by waste agents who burn a lot of fuel and add mileages to convey waste to disposal facility. The solution to this problem is the reorganization of the town to reflect proper town planning for old municipality and the establishment of a well-planned city for new towns that are springing up from the suburbs. This involves the establishment of direct road link to waste facilities that are far away from human habitation.

8. Conclusion

Based on the study carried out, it is important to note that disposal of refuse in mangrove forest has a boomerang effect. This is because heavy metals and other pollutants from the waste percolate into ground water to contaminate the drinking water source in nearby communities. It is also known that mangroves serve as the food basket of the sea, and any pollutant that enters into it will be redistributed back to man through the food chain. Although, mangroves are resilient, and can withstand pollutants but their resilience should not be taken for granted because it can lead to ripple effect that will eventually affect mans' health especially in the aspect of increase in heavy metals up the food chain through biomagnification.

Author details

Aroloye O. Numbere
Department of Animal and Environmental Biology, University of Port Harcourt, Choba, Nigeria

*Address all correspondence to: aroloyen@yahoo.com

References

[1] Banfield EC, Wilson JQ. City Politics. Cambridge: Harvard University Press and MIT Press; 1963

[2] Boesch DF. Diversity, stability and response to human disturbance in estuarine systems. In: Proceedings of the First International Congress of Ecology, Wageningen, The Netherlands. 1974. pp. 109-114

[3] Gunderson LH, Holling CS, Pritchard L Jr, Peterson GD. Resilience of large-scale resource systems. In: Gunderson LH, Pritchard L Jr, editors. Resilience and the Behavior of Large-Scale Systems, SCOPE Series. Vol. 60. Washington: Island Press; 2002. pp. 3-20

[4] Numbere AO. Impact of Hydrocarbon Pollution on the Mangrove Ecosystem of the Niger River Delta, Nigeria [dissertation]. Saint Louis University; 2014

[5] Numbere AO, Camilo GR. Mangrove leaf litter decomposition under mangrove forest stands with different levels of pollution in the Niger River Delta, Nigeria. African Journal of Ecology. 2016;**55**:162-167

[6] Numbere AO. Comparison of microbial and heavy metal contents in soils and roots under mangrove forest stands with different levels of pollution in the Niger Delta, Nigeria. American Journal of Applied Sciences. 2017;**15**(2):132-140

[7] Lacerda LD, Machado W, Moscatelli M. Use of mangroves in landfill management. GLOMIS Electronic Journal. 2000;**1**(1):1

[8] Lacerda LD, Carralho CE, Tanizaki KF, Ovalle ARC, Rezende CE. The biogeochemistry and trace metals distribution of mangrove rhizophores. Biotropica. 1993;**25**:252-257

[9] NPC. Current population of Nigeria. 2017. Available from: www.npc.org [Accessed: May 26, 2017]

[10] Abrahamson DB. Urban planning in China: Continuity and change: What the future holds may surprise you. Journal of the American Planning Society. 2006;**72**:2

[11] Measham TG, Preston BL, Smith TF, et al. Adapting to climate change through local municipal planning: Barriers and challenges. Mitigation and Adaptation Strategies for Global Change. 2011;**16**:889. DOI: 10.1007/s11027-011-9301-2

[12] Guerrero LA, Maas G, Hogland W. Solid waste management challenges for cities in developing countries. Waste Management. 2013;**33**(1):220-232

[13] Sharholy M, Ahmad K, Mahmood G, Trivedi RC. Municipal solid waste management in Indian cities–a review. Waste Management. 2008;**28**(2): 459-467

[14] Kennedy C, Pincetl S, Bunje P. The study of urban metabolism and its applications to urban planning and design. Environmental Pollution. 2011;**159**(8):1965-1973

[15] Sumathi VR, Natesan U, Sarkar C. GIS-based approach for optimized siting of municipal solid waste landfill. Waste Management. 2008;**28**(11):2146-2160

[16] Saeed MO, Hassan MN, Mujeebu MA. Assessment of municipal solid waste generation and recyclable materials potential in Kuala Lumpur, Malaysia. Waste Management. 2009;**29**(7):2209-2213

[17] Al-Salem SM, Lettieri P, Baeyens J. Recycling and recovery routes of plastic solid waste (PSW):

A review. Waste Management. 2009;**29**(10):2625-2643

[18] Laner D, Crest M, Scharff H, Morris JW, Barlaz MA. A review of approaches for the long-term management of municipal solid waste landfills. Waste Management. 2012;**32**(3):498-512

[19] Ghiani G, Laganà D, Manni E, Musmanno R, Vigo D. Operations research in solid waste management: A survey of strategic and tactical issues. Computers & Operations Research. 2014;**44**:22-32

[20] Zhang DQ, Tan SK, Gersberg RM. Municipal solid waste management in China: Status, problems and challenges. Journal of Environmental Management. 2010;**91**(8):1623-1633

[21] Scarlat N, Motola V, Dallemand JF, Monforti-Ferrario F, Mofor L. Evaluation of energy potential of municipal solid waste from African urban areas. Renewable and Sustainable Energy Reviews. 2015;**50**:1269-1286

Chapter 11

Improper Disposal of Household Hazardous Waste: Landfill/Municipal Wastewater Treatment Plant

Elsayed Elbeshbishy and Frances Okoye

Abstract

Household hazardous waste (HHW) is not always separated for proper handling before disposal. When disposed improperly to landfills and municipal treatment plants, these products can have significant impact on the environment. Although HHW is a small portion of municipal solid waste, the presence of HHW in solid management facilities that are not equipped to handle them can have problematic effects, resulting in environmental pollution, damage to facilities, and even injury to workers. In many countries, HHW is not subject to legislation unless separated from other household waste because of its small percentage and the challenge in enforcement. In addition, there is no standard definition of HHW globally; therefore, what constitutes to HHW in one country may not be in another. Government legislation and schemes such as Extended Producer Responsibility play a vital role in encouraging proper disposal among consumers, especially when they are convenient and accessible. In this chapter, hazardous household products in different countries are considered along with common improper and acceptable disposal methods. Furthermore, the impacts of improper disposal on the environment are explored with an emphasis on landfill leachate and wastewater treatment plant effluent. Finally, current legislation and programs that encourage proper disposal are discussed.

Keywords: household hazardous waste, disposal, environmental impact, landfill, wastewater treatment plant, legislation, schemes

1. Introduction

Household waste is something that is common among most, if not all, living residences. Like any industrial facilities that handle potentially hazardous materials, households too dispose and use hazardous substances. The chemical complexion in the waste substances makes it so if disposed improperly, it could ignite, explode, poison, or corrode. Household hazardous waste (HHW) becomes what it is once thrown away. Methods of the waste being improperly disposed is pouring the substance down the drain, into storm sewers, on the ground, and throwing it in among the trash. It may not be obvious that these substances, once disposed, will be a danger, but particular varieties of HHWs have the prospective to:

- cause somatic injury to sanitation workers;
- if poured down drains or toilets, adulterate septic tanks or wastewater treatment systems;
- pollute—if poured down storm sewers—bodies of water;
- become a danger to young or unknowing children and pets if left open in the house;
- contaminate ground and/or surface water that is used as a way of obtaining drinking water, if directed to exposed landfills.

A big problem that occurs/can occur through improper disposal of HHW would be the deconstruction that the sewage treatment plants are able to obtain. These plants are not able to deconstruct HHW compounds that people would drain or flush, which will end up traveling into lakes and rivers, unprocessed. As a result, one of the main releasers of dioxins and furans was from sewage systems. The substances proved to threaten human health due to the fact that they were highly carcinogenic. Other than the fact of the carcinogenic dangers, interference with the treatments plants could transpire. The toxins that would be processed could poison the microorganisms in the biological process. That would bring us to the position where our water systems would be more susceptible to harmful contaminants.

As a given, hazardous waste is poisonous to all life forms, exposure of such hazardous substances to any living organism (plants and animals) could devitalize it. As a consequence, to the environment, hazardous waste could diminish natural resources and be contaminating to humans. Giving the young/fetuses, whether human or animal, exposure to these hazards would be substantially dangerous, as they are in a process or rapid growth. Introduction to chemicals for the living body would also interfere with biological structure, causing malfunction of organs and limbs.

In addition to the effects to the human and animal bodies, hazardous waste would hinder plant growth. The impeding of plants that are of much use to the human race through manufacturing and consumption would affect our habitat. If the plants were slowly changing, for the worse, it would affect the animals that are needed for food, farm work, and would cause a whole new era of extinction.

If our plant growth can affect our way of living easily, dumping the HHW into landfills gives us a much bigger problem. Landfills that are improperly maintained are major problem; even if they seem to be isolated from any contact, they can contaminate the environment around them. These landfills produce foul-smelling and toxin gases. Along with the gases and toxins, landfills generate leachate, which can travel to our water sources of lakes, rivers, and oceans. This would dig us into a deeper problem of both environmental and human existences. Thereby, leaving HHW unattended and improperly disposed could potentially destroy the ecosystem.

2. Household hazardous waste (HHW)

Separate management of HHW from nonhazardous waste is rare. It is estimated that in countries within the Organization for Economic Cooperation and

Development (OECD), household waste contributes to 67% of 540 million tons of municipal solid waste (MSW). The estimated amount of HHW varies considerably due to an unclear definition of what constitutes to household waste as opposed to MSW. In the USA, for the Environmental Protection Agency (USEPA), household products that contain corrosive, toxic, ignitable, or reactive ingredients are considered to be HHW. In general, the HHW is a solid, semisolid, or nonaqueous liquid that can cause or significantly contribute to potential hazard to human health or environment when it is improperly treated, stored, transported, disposed of, or otherwise managed. The portion of HHW in MSW has been estimated to be from less than 0.01–3.4% in several studies. The large variability is due to lack of standard definition as to what constitutes HHW, variability in generation, variability in weighing methods, and limited sample size. Nevertheless, 1% by weight is widely accepted as the fraction of HHW in MSW. Because of this small percentage of HHW produced, households are not practically considered to be hazardous waste producers [1]. While HHW represents a relatively small proportion of current urban solid residues, it is the most toxic part of the waste stream.

HHWs in the household waste are often excluded from management as hazardous waste unless collected separately. However, if these waste materials were generated industrially or commercially, they would be subject to strict disposal guidelines. As a result, HHWs are handled the same way as nonhazardous material with no specific regulation or monitoring. Of recent, this mismanagement constitutes a greater problem as the waste stream not only increases in amount but also becomes more diverse with the introduction of more products into the consumer market.

At the source or point of generation, HHW can be placed in the garbage, down the drain, dumped on the ground, or diverted for reuse, energy recovery, or recycle. No matter where HHW is disposed, due to its toxicity as well as municipal treatment facilities that are not equipped to deal with hazardous material, improper management can adversely impact the quality of the environment:

- Contaminate ground water bodies.
- Contaminate surface water bodies.
- Pollute air.
- Affect the human health (children and pets if left around the house, cause physical injury to sanitation workers).

On the other hand, in many third world countries, solid waste management facilities are underdeveloped and sometimes nonexistent. The United Nations reported that between 20 and 80% of all household waste that is generated is often dumped in open spaces, water bodies, drains, and burnt or buried. This creates unsanitary environments leading to health hazards. The portion of HHW in household waste generated by developing countries is much less than in developed countries. The small amount of HHW produced as well as unavailability of funds to direct toward implementing sound practices for waste management has led the United Nations Environment Program to suggest HHW with MSW for disposal in landfills [2]. Regardless of the development level of the country, proper management of HHW can be achieved by understanding the environmental and societal impact of poor practices, HHW contaminants, government legislations, and well-developed schemes.

3. Potential risks of improper disposal of HHW

Unavailable facilities for proper HHW management discourage even their voluntary participation. While the products in the HHW list vary from country to country, below are categories that represent majority if not all products that can be classified as HHW:

- Photochemicals
- Pesticides
- Mercury-containing wastes
- CFC-containing equipment
- Nonedible oil and fat
- Paints, inks, resins, and adhesives
- Detergents
- Pharmaceuticals
- Batteries
- Waste electrical and electronic equipment
- Wood preservatives
- Aerosols
- Personal care products

The risks that a hazardous product poses to the environment depend on certain characteristics of the toxic compounds:

- Solubility
- Mobility
- Persistence
- Degradability
- Toxicity to nonhuman target species
- Potential for penetrating landfill liners
- Potential to be broken down in wastewater treatment system

HHW is likely to be disposed of improperly because residents do not always understand the level, effect, and potential impact of toxicity in the products that they use. In **Figure 1**, the disposal trend of households in the UK is presented after a survey with 400 respondents was carried out. One can observe that the predominant method for disposing HHW in households is into the garbage in spite of the toxicity level. A large portion of photochemicals and pharmaceuticals are discarded down the drain with little regard for the compounds that they contain and the consequences for this mode of disposal [4].

Since information about the impact of HHW on the environment is not exhaustive and data relating to disposal are not well known, the potential impact of each of these products in the environment and health is considered as well as the amount that is approximately generated by households where available.

On the other hand, **Figure 2** shows a similar study conducted by Statistics Canada in 2009 with over 3800 respondents. While the garbage is still a significant disposal route for HHW, more households reported utilizing drop-off centers and returning products to suppliers and retailers [5].

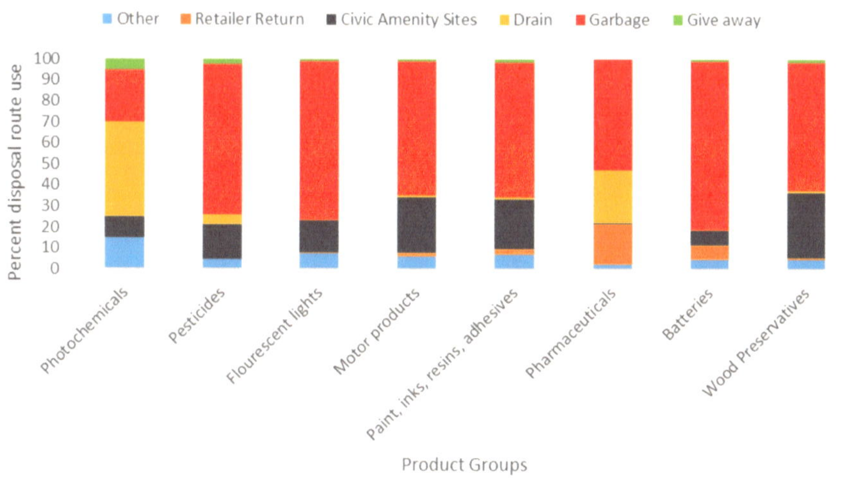

Figure 1.
Usual HHW disposal regime of UK households [3].

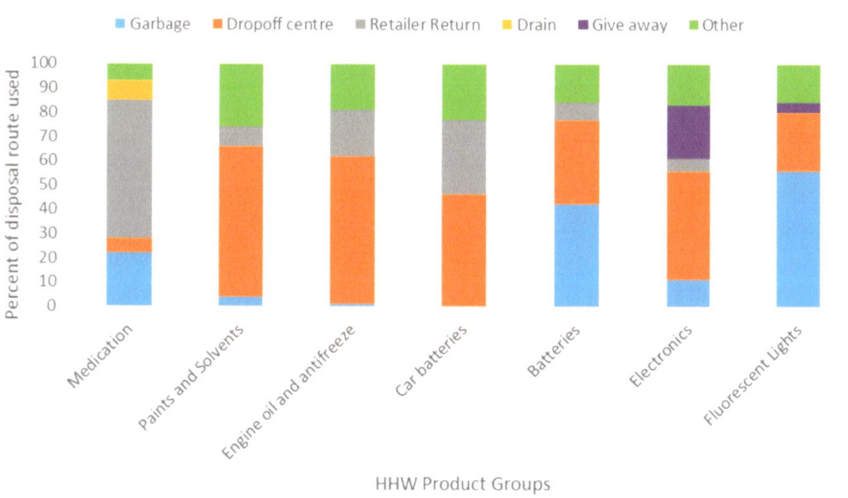

Figure 2.
Usual disposal routes in Canadian households (source: Statistics Canada).

3.1 Photochemicals

These are liquid chemicals used in home developing and printing. Many of the ingredients in these products are toxic solvents and are predominantly disposed of in sewers. The unused portions of these chemicals are hazardous, but also the packaging can be problematic as it can contain some of the chemicals, which end up in the landfill and thus contaminate both soil and groundwater because these chemicals can penetrate the liners transporting to the groundwater and might end up to the surface water through the movement of groundwater. While the amount entering the sewers cannot be estimated, the packaging in the UK is estimated to be about 270 tons/year, most of which will end up in the landfills [6].

3.2 Pesticides

Rapid growth in pesticide use has been observed, and this suggests a proportional increase in the amount that is being disposed of. According to the UK Pesticide Safety Directorate, many of the active compounds have been observed in landfill leachate of which research shows that they pose carcinogenic and endocrine disruptive risks [6]. On the other hand, incineration of pesticides is acceptable, provided that they do not contain mercury or arsenic. In Belgium, around 80% of waste pesticides are collected and incinerated [7].

3.3 Mercury-containing wastes

Household products that contain mercury include fluorescent bulbs, stockpiled paint, dental amalgam, thermometers, and barometers. Of these, fluorescent bulbs contribute the highest amount of mercury waste. However, as the use of these is reported to have better energy and environmental impacts than regular light bulbs, they are so encouraged [8]. Improper disposal of fluorescent bulbs is where the risks lie. In the UK, it is estimated that 80 million are disposed of each year, of which only a small portion are recycled or processed for mercury extraction. In Brazil, lamps containing mercury contribute 1000 kg of mercury disposed of per year. Mercury exposure poses some health risks such as genetic damage and neurotoxicity damaging the kidney, liver, and central nervous system [6].

3.4 Chlorofluorocarbon (CFC)-containing equipment

Refrigeration and air-conditioning appliances/equipment may contain chlorofluorocarbon (CFC) and hydrochlorofluorocarbon (HCFC) refrigerant. CFCs and HCFCs are ozone-depleting substances (ODS). If they released to the environment, they will destroy the protective ozone layer above the earth and potent greenhouse gases, contributing to global climate change. Examples of these types of equipment include motor vehicle and motor vehicle-like air conditioners, central and room air-conditioning units, refrigerators, freezers, chillers, drinking water coolers, dehumidifiers, research equipment, vending machines, etc. Manufacturing of such freezers and refrigerators has been phased out with the CFC component being replaced. However, disposal of these is still ongoing because of their 8- to 12-year life span leaving 4500 tons of CFC in the UK to be safely disposed of. Disposal of equipment that contains ODS is regulated in the EU by the WEEE directive where separate collection is mandatory [6].

3.5 Nonedible oil and fat

Nonedible oil and fat constitutes to about 15% of HHW in the UK. The waste section comprises mineral oils that often contain additives, which make it hazardous. While they are sometimes collected and rerefined or burned for energy, a significant portion is disposed of by end users down the drain or via oil filters and end up in the landfills. There, the oils can disrupt artificial landfill liners. Preferably, the steel component oil filters can be recycled after the oil is pressed for recovery and processed into fuel by companies [7]. Other examples are maintenance lubricants and greases for vehicles, which contain solvents and hydrocarbons that can be just as harmful.

3.6 Paints, inks, resins, and adhesives

Disposal of paints is the most significant in this category with the solvent-based paints posing the higher risk. In the UK, paints contribute to 17% of the total HHW with large quantities ending up in the sewers or mixed with MSW. However, schemes developed by local charities exist to collect unwanted paint and redistribute them at no charge. This scheme is limited by the quality of paint that can be used and quantity that can be accepted in any given location. Collected paint needs to meet certain criteria to be acceptable for redistribution such as age or packaging. Barely, 1% of the available excess paint is collected due to a small number of collection points [6, 7].

3.7 Detergents

The use of detergents in household is widespread. It has been reported that 5–20% of the phosphate that is found in surface and ground water in northern Europe originates from detergent use. However, not all detergents are classified as hazardous, but those containing acids, bases, and chlorinated solvents are of particular concern [7]. In addition, the biodegradability, the aquatic toxicity, endocrine disruptiveness of the surfactants, and other ingredients in the detergent contribute to its classification as hazardous. When combined, some compounds in detergents can release fumes that affect the eyes and mucous membranes, leading to respiratory failure and death after prolonged exposure [6].

3.8 Pharmaceuticals

In the US, all over-the-counter (OTC) medicines are regarded as hazardous. However, in the EU, only those that are cytotoxic are classified as hazardous. Consumer disposal is not particularly regulated, as it would be problematic, but also due to the relatively low toxicity. As a result of the inability for wastewater treatment plants to remove pharmaceuticals from the waste stream, in many countries, they are now regarded as water contaminants. This is because they eventually make their way into drinking water supplies. They are transferred to sewage sludge during treatment, which is then applied to agricultural land or sent to a landfill [6].

3.9 Batteries

Primary, lead-acid, and nickel-cadmium batteries are those that fall into this category in HHW making up 6–14% of the HHW in the UK. Mercury in consumer batteries has been banned in Europe and many states in the US. However, many unregulated countries still use batteries containing significant concentrations or mercury, which often ends up in landfills. When buried in landfills, the casing of dry cell batteries can degrade and release heavy metals [9]. Most rechargeable batteries are used in consumer devices and nickel cadmium batteries. In the EU, these types of batteries must be easily removed from electronic devices, and separate collection for recycling is encouraged. However, these end up in MSW where recycling facilities are not well established because it is not mandatory. Cadmium is known to cause health effects like kidney damage. Lead-acid batteries comprise those found in vehicles, or smaller batteries in fire and security alarms. The recycling program for lead-acid batteries in the UK is well established, and 85% collection of the automotive variety has been recorded. However, the batteries from the alarms and from some battery changes carried out at home still end up in MSW. Lead acts as a chronic and acute neurotoxin affecting the kidney [6].

3.10 Waste electrical and electronic equipment (WEEE)

For many years, home electric and electronic equipment has been disposed of in landfills along with their hazardous components. The amount that is being disposed continues to grow as consumer interest in current devices keeps increasing, which leads to discarding of obsolete electronics. WEEEs often have toxic compounds such that special handling is a requirement [8]. Many countries have prohibited the disposal of WEEE in landfills because of the toxicity and the strain of such large quantities of waste on the landfills. In the EU, this group of equipment is regulated under the WEEE directive such that they are collected and treated as hazardous waste. The directive also lists the substances that should be removed and collected from WEEE. Restrictions have also been placed on the use of certain materials in the manufacture of newer equipment [6].

3.11 Wood preservatives

There are three types of treatments that are used to preserve wood, all of which can cause the treated wood to be hazardous, as they have hazardous properties. The types are tar oils, organic solvent-based, and water-based formulations. Creosote, an aquatic contaminant, is often used in tar oils. It is known to be a skin irritant, which causes photosensitivity and skin tumors following long exposure. Tributyltin is an example of organic solvent-based compound that is strictly regulated. A hazardous water-based substance is copper-chrome arsenate (CCA), which contains concentrations of heavy metals that have large health and environmental risks [6]. Arsenate is a priority carcinogenic contaminant of waste, which easily leaches in a landfill and can volatilize during incineration. Landfilling is not acceptable for disposal, and specialized air pollution control equipment is required for incineration [7].

3.12 Aerosols

Aerosols are a large portion of HHW making up 26% of the HHW in the UK. In the past, CFC was widely used in the production of aerosols. However, CFC has been replaced with alternative propellants and solvents, which contribute significantly to the content in HHW. These replacements are often flammable and explosive. Exposure to aerosols can lead to nausea, skin, and throat irritation [6].

3.13 Personal care products

The harmful nature of PCP has been supported by the discovery of certain long-term effects on health and the environment. While most PCP will end up in the sewers, unused products are stockpiled and end up in MSW.

It is important to understand the fate of compounds in HHW when mixed with MSW for disposal. This has led to stricter disposal regulations in many developed countries to improve HHW management [6].

4. Environmental impact

Improper disposal of HHW eventually leads to the presence of hazardous contaminants in the environment. All the facilities that are used to manage discarded HHW are in direct contact with environment media, air surface water, groundwater, and soil (**Figure 3**). These media are in constant contact with each other. As a result, when facilities cannot adequately break down hazardous compounds in HHW, the immediate environment is at risk.

The contaminants enter the water cycle via groundwater or lakes, rivers, and streams traveling through the cycle [10] via different paths:

- Precipitation from the atmosphere
- Percolation through the soil
- Direct disposal from a wastewater treatment plant (WWTP) into a surface water body
- Residents pouring liquids down the stormwater drain that empties into a lake

In addition, toxic gases from HHW can be emitted into the air from the hazardous compounds that are used in producing them during controlled incineration or sometimes, uncontrolled fires [8].

4.1 Landfills

Landfills can be the most economic way for waste management, especially in countries like Canada with large open spaces. However, poorly managed landfills have the potential of causing a number of environmental issues such as contamination of groundwater or aquifers or soil contamination. Modern landfills are not just holes in the ground to be packed with garbage. They can be considered as highly engineered contaminated systems. A modern landfill uses a number of technologies to ensure that the wastes are properly managed to avoid environmental pollution (e.g., ground water contamination, gas emission). **Figure 4** shows schematic of a modern landfill process. Advanced protective liners (both natural and manufactured) are typically used to isolate the waste and leachate from leaking into the surrounding ground or ground water. Single, composite, or double liners can be used depending on the nature of the waste materials being deposited (see **Figure 5**). At minimum, a composite liner should be used for hazardous waste landfill facilities. However, landfills are not usually engineered to handle toxic compounds from HHW [9]. Hazardous liquid waste can be transported from a landfill into the environment if there are no barriers. Leachate that has been contaminated with hazardous material (soluble or insoluble) may destroy synthetic liners and render

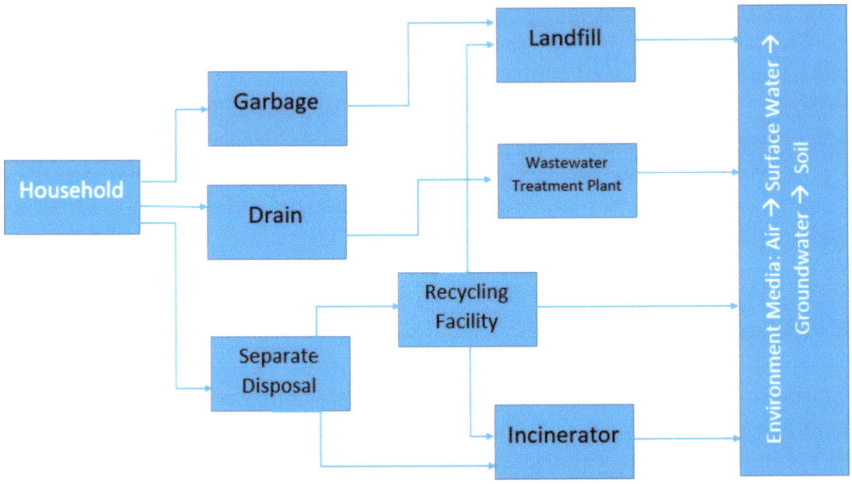

Figure 3.
Improper disposal path of HHW from household to environment.

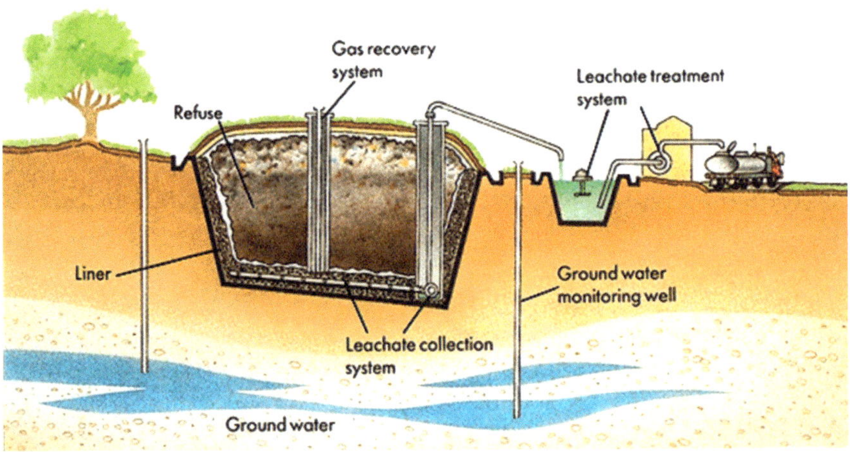

Figure 4.
Schematic of modern landfill process (source: www.oocities.org).

existing barriers ineffective, and thus, the hazardous waste comes in contact with the soil. Its fate is determined by the characteristics of the soil such as porosity, geological factors, and the contaminant like viscosity. The contaminant may percolate downward and affect the groundwater or spread and contaminate surrounding area [10]. Even if the leachate is collected, the treatment plants are not usually equipped to remove hazardous contaminants and end up releasing them into water bodies [11].

In addition, the conditions of the landfill such as the air and moisture content can affect the fate of hazardous contaminants such as the rate of degradation or violent reactions [10]. For example, phthalic acid esters (PAEs) are used as plasticizers that are used in furniture, clothes, food packaging, etc., which are items that will invariably end up in the landfill. While readily degradable under aerobic conditions, those that are found in the landfill environment tend to retard biodegradability of PAEs. When the environmental impact of PAE in a landfill in China was studied, it was discovered the more complex congeners were found absorbed in deeper soils and in the groundwater [12].

Certain volatile organic compounds can be partially degraded and are readily absorbed by MSW in a landfill rather than volatilize. The moisture in the leachate enhances this process. Leachate-containing toxic compounds can be detoxified faster by recirculation within the landfill, which reduces the potential for leakage from the landfill liner. HHWs contribute volatile organic compounds (VOCs) to landfill gases such as benzene, methylene chloride, trichloroethylene, vinyl chloride, etc. VOCs from landfill gases contaminate off-site groundwater through migration [11].

4.2 Incinerators

The quality of air emissions and ash residue is as a result of the fuel being incinerated. Incinerators usually have pollution control devices; however, some of the components that are found in HHW can pose a challenge to be captured. For example, mercury found in dry cell batteries, fluorescent light bulbs, and old paint can be converted to gaseous form and be emitted from the stack. Even the use of air treatment technologies can only remove 75–85%. Once it becomes in the atmosphere, mercury can be solubilized by rain and end up in water bodies. Other contaminants such as hydrogen sulfide and carbon monoxide that enter the atmosphere as gases may react with other compounds to become even more hazardous or remain in the atmosphere if stable, causing damage. Also, toxic metals have been found in the fly ash residue of incinerated MSW containing HHW. Damaging explosions have been reported due to a flammable liquid container being heated, which can lead to a few hours to few years of lost work time [11].

4.3 Wastewater treatment plant

Hazardous material dumped down the drain will end up in the on-site septic system or wastewater treatment plant depending on which system is employed. HHW can enter into wastewater treatment systems through its intended use or as a disposal method. Local governments usually prohibit disposal of HHW into stormwater drains. Recommended disposal may depend on the product and the industry. Some may be dumped down the drain with lots of running water, while others should be kept for collection [11].

Conventional wastewater treatment plants combined physical, chemical, and biological treatment methods depending on the nature of the pollutants and desired level of removal. Modern wastewater treatment process consists of four levels, including preliminary, primary, secondary, tertiary, or advanced treatment, in addition to the solid waste management. Preliminary and primary treatments are mainly physical/mechanical (screening and gravity settling), while secondary and tertiary treatments use combination of biological, physical, and chemical treatment process (**Figure 6**). Preliminary treatment removes larger inorganic materials and floating particles, primary treatment removes a major portion (50–60%) of suspended solids from raw wastewater, and secondary treatment process removes organic matters and suspended solids. Secondary treatment usually consists of biological treatment of wastewater. Most of the WWTPs use aerobic activated sludge process for secondary treatment. The objectives of secondary treatment are to reduce BOD and SS of the effluent to an acceptable level according to the discharge regulation. In some cases, nutrient removal may be also an objective of secondary treatment. Biological treatment processes rely upon the ability of the organisms to utilize the contaminants as substrates and results in the generation of new biomass and biodegradation by-products.

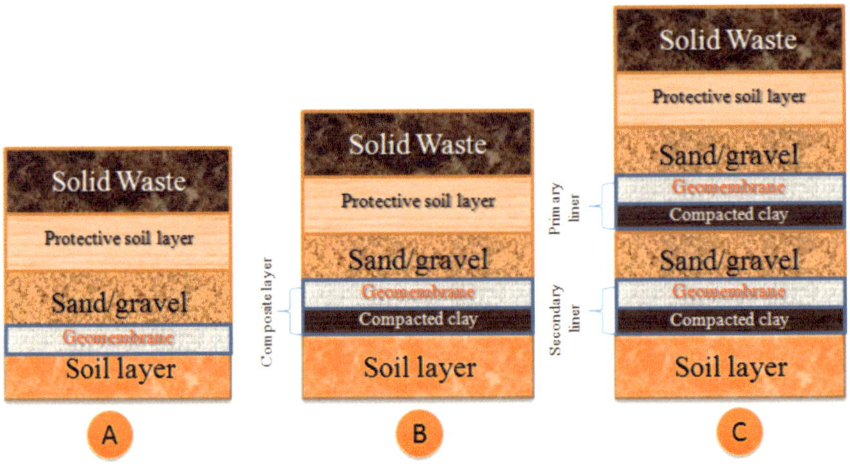

Figure 5.
(A) Single liner, (B) composite liner, and (C) double liner system.

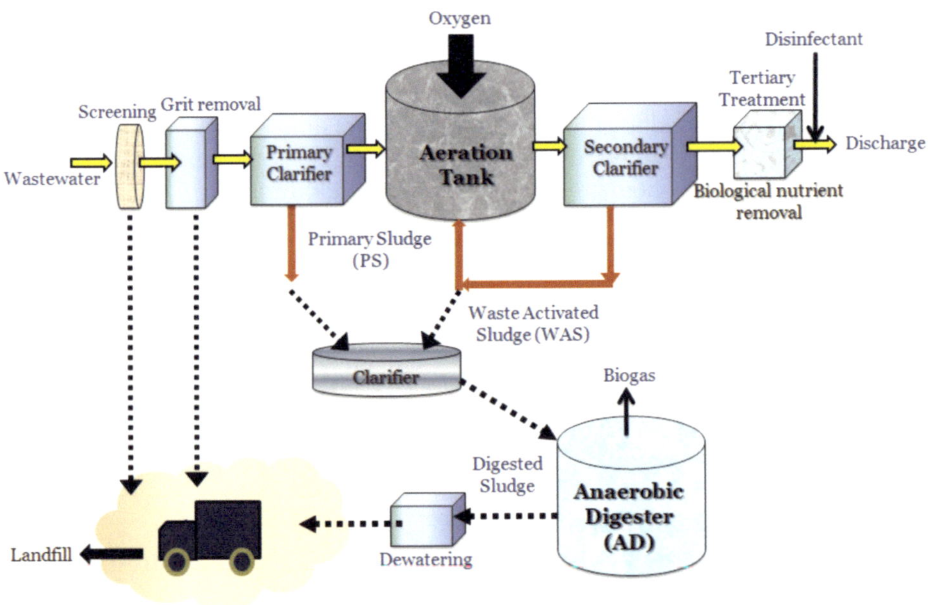

Figure 6.
Typical municipal wastewater treatment process.

Lye and bleach found in cleaning products and other hazardous components can hinder the bacteria that are utilized in the biological treatment processes and will significantly affect the process efficiency. This will cause wastewater to pass through the system without treatment and ultimately will reach the groundwater and/or surface water [10]. This can contaminate aquatic life; nitrates, and phosphates can cause eutrophication (algal bloom), leading to the use of more herbicides for control.

Excess loading of nutrients like nitrates and phosphates results in the uncontrolled growth of phytoplanktons and macrophytes. The growth and subsequent death of these organisms form a greenish slime layer at the surface of water bodies. This slime layer reduces the amount of sunlight that can penetrate through and the oxygen that can be replenished into the water. In addition, the excessive growth causes high competition for resources among aquatic organisms and death such that the biodiversity in the water body may be severely affected over time. This is the water pollution phenomenon known as eutrophication. Aside from the negative effects on water esthetics, eutrophication can hamper recreation activities, navigation, and aquatic life [13].

On the other hand, heavy metals are toxic, persistent, and mobile and tend to accumulate. They generally have very low acceptable concentrations in drinking water standard. In WWTP, low-concentration volatile solvents can evaporate from the aeration tank and become air pollutants. However, high concentrations, acids, bases, poisons, and solvents can affect the WWTP workers' safety and effluent quality and contaminate the sludge. Even if the wastewater flow does not contain HHW, leachate from landfills and combined sewer flow can introduce contaminants from pesticides and motor oil, which originate in the households of which even a small amount of pesticide concentration can cause a WWTP to fail toxicity test [11].

4.4 Recycling centers

Majority of the e-waste collected in the US and other developed countries end up in developing countries in Asia and Africa, which often have less than adequate concern for the environmental impacts of the primitive recycling activities that are conducted. Illegal e-waste recycling activities in Guiyu, China, have led to the release of hazardous chemicals into the environment. Harmful concentrations of heavy metals and compounds such as polybrominated diphenyl ethers (PBDEs) were reported in local children and workers of the recycling facilities likely due to open dumping activities that contaminated the soil and river sediments. Polychlorinated biphenyls (PCBs) released during manual dismantling of electronics and from open combustion of the waste material resulted in the presence of significant concentrations in the local residents as a result of bioaccumulation in fish and inhalation [14, 15].

5. Proper disposal methods

5.1 Source separation

Proper disposal of HHW starts with differentiating between hazardous household products and nonhazardous waste products. Mixing of household waste at the source must be addressed and banned. By collecting similar HHWs together, they can be more efficiently managed with regard to environmental safety, human health, and costs. When separated, arsenic-treated wood can be incinerated using proper pollution control technologies reducing any form of carcinogenic environmental impact, which may otherwise be present if it had been landfilled. Even more popular in developed countries is the separation and collection of cleaning products and pesticides. Majority of which can be incinerated according to best practices unless they contain mercury [7].

5.2 Recycling and repurposing

Some HHW products can be of value as they can be recycled for a different purpose or may contain material, which can be extracted for use in manufacturing other products, as in the case of antifreeze, which can be repurposed as an engine coolant. Another example is waste motor oil, which can be refined as lubricating oil

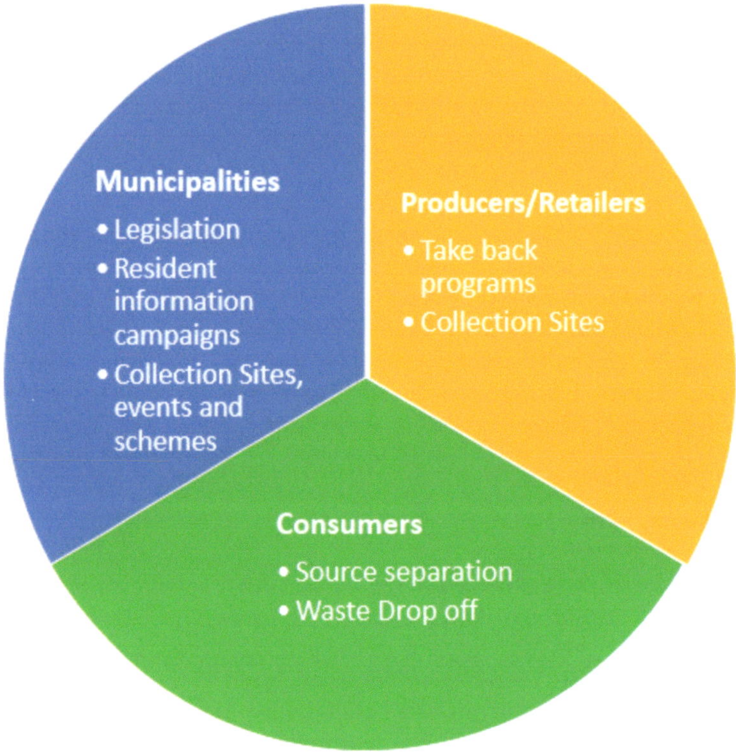

Figure 7.
Collaboration between municipalities, producers, and consumers for proper HHW disposal.

or processed as low-grade fuel oil. Lead-acid batteries contain lead, which can be extracted to produce new batteries. Dry cell batteries, on the other hand, contain many different heavy metals, which may pose a problem for extraction. By collecting a significant amount separately, they can be disposed of more cautiously as hazardous waste [10]. Many EU states collect and recycle fluorescent tubes; however, in Germany, all the components of fluorescent tubes aside from the fluorescent powder have been reported to be reused [7].

5.3 Give out

Rather than discard surplus products in the garbage or down the drain, items such as paint and wood preservatives can be given out to those who require them when in good condition. Charities that facilitate these have been established in certain countries.

6. Legislation and policies

For any management system to be successful, efforts from the municipalities, manufacturers, and residents must be combined. Legislation that assigns responsibility for hazardous components and clarifies handling requirements of household hazardous products encourages manufacturers to consider sustainable methods of recycling waste from their products. Collection programs and proper management schemes fostered by municipalities and industries working together can reduce the amount of HHW that is discarded dangerously. At the core of these programs is the voluntary source separation by residents within their households. The participation of municipalities, manufacturers, and retailers is also required for these programs to be successful at HHW management (**Figure** 7) [1].

6.1 European Union

The member states of the European Union (EU) are subject to the Waste Framework Directive (WFD) or the Directive 2008/98/EC of the European Parliament concerning general requirements for waste management. Established in 1975, this directive has been substantially amended with the latest revision provided in 2008. HHW is covered in article 20 of this directive, and as with previous directives, it is excluded from the definition of hazardous waste, while it is mixed with other types of household waste. The exclusion also applies when HHW has been separated from mixed household waste and remains until it has been collected properly. Under this legislation, there is no guidance to the management of HHW or legal obligation to the house owners [1, 16].

Directives exist for specific categories of hazardous waste. These documents provide some direction for member states on collection and disposal of the waste and encourage the education of householders on the importance of separating HHW from mixed municipal waste and of the collection and recycling programs that are available to them. The categories include waste from electrical and electronic equipment (WEEE), batteries and accumulators, and waste oils.

Waste oils are covered under the WFD directly in article 21. About 3 million tons of waste oil need to be managed annually in the EU that can severely damage soil and water. The directive prohibits any type of disposal that may adversely affect the environment and human health, discourages mixing of different types of waste oils, and encourages separate collection. The directives for 'batteries and accumulators' and WEEE call for accessible and free collection points and requires producers and distributors to take back waste batteries, accumulators and electrical and electronic equipment (EEE). However, while the disposal of industrial and automotive batteries and accumulators in landfills and incinerators are prohibited, no such legislation is put forth for household batteries. On the other hand, disposal of WEEE is prohibited until proper treatment has been carried out [17].

6.2 North America

HHW in the US is regulated under Subtitle D of the Resource Conservation and Recovery Act as solid waste. It is excluded from hazardous waste, provided that it is material from a permanent or temporary residence [18]. However, since solid waste is regulated by the state and local authorities, some states have more stringent regulations for the management of HHW.

An example is Hawaii. Many of the items that are on the federal list as HHW are the same in Hawaii. However, lamps that contain lead and/or mercury and lead-acid batteries are managed more strictly. In addition, an electronic bill was passed that required computer manufacturers to establish recycling programs. There is also a prohibition on placing motor oil on the ground, in the drainage ways, in sewers, or into water bodies [19].

In Canada, the disposal of solid waste falls under the care of the municipalities with the provinces monitoring operations. While HHW is limited to paint, aerosols, solvents, pesticides, and other products containing hazardous properties, the Waste Diversion Act sets the requirements and guidelines for management of HHW, WEEE, and waste oils in Ontario. Under this act, manufacturers are financially responsible for HHW program, which was developed in 2006 to manage waste from their products. Similarly, WEEE management is mandatorily funded by industry though some retailers charge consumers an environmental fee at the time of purchase of electronic equipment. However, recycling of mercury-containing lamps is voluntary for consumers [20].

6.3 Schemes and programs

Municipalities may establish frequent HHW curb-side pickup as part of the general waste collection program. While convenient for households, this mode can be expensive and time-consuming for waste management authorities [1, 19]. Other options include less frequent collection such as biannually, residents requesting special wastes pick up or personal drop-off at central locations. Such programs for proper disposal and recycling are well established in many countries including the US, Australia, Germany, Denmark, and Sweden [1, 7].

The Extended Producer Responsibility (EPR) is a government policy approach that places the main responsibility of managing a product on the producers or manufacturers. The EU's WEEE directive and Hawaii's computer recycling program are primary examples of such a legislative approach. Companies within an industry can collaborate to develop initiatives for handling waste from their products. The Rechargeable Battery Recycling Corporation (RBRC) is a company that was created by the efforts of battery manufacturers in North America. RBRC is responsible for collecting and processing certain types of batteries in order to extract metals that can be used in manufacturing new batteries [11].

Very similar to the EPR is the Product Stewardship (PS) approach. The manufacturers, retailers, and consumers share the responsibility for the end of life management of a product. The EU has programs similar to these for the management of pesticides and air fresheners [11]. The US also has a well-established PS system and enforces these programs in some states through laws, subsidies, fees, and mandatory take-backs [1]. The retail take-back system provides a setting for retailers to collect waste materials from consumers whether for exchange or refunds. It is particularly attractive because retailers tend to be within reach and more convenient for consumers. However, there is the potential for such a program to place financial burden on the retailers due to handling and storage requirements. In North America, it has been used successfully for the management of all kinds of waste including automotive batteries, fluorescent lamps, mercury thermostats, etc. Likewise, Japan has a take-back program for home appliances, but it is mandatory and requires consumers to pay the retailer for the waste handling [3].

7. Conclusions

In this chapter, the adverse impacts of improper disposal of HHW on the environment were discussed. Improper disposal of HHW introduces harmful compounds, which cannot be removed by treatment facilities into the environment, and these chemicals end up in human, animal, and plant tissues. What constitutes to inadequate disposal varies from pouring down the sink or drainage, dumping in the garbage or even out on the ground outdoors. Even when proper disposal routes are provided by municipalities such as drop-off centers are available, many classes of HHW are still disposed in the garbage. Public education, source separation, and recycling are key strategies to reducing the quantity of HHW stream into municipal facilities and by extension of the environment. The success of these strategies for HHW disposal requires voluntary action from residents, legislation from governments mandating manufacturers to take better responsibility, and schemes that make proper disposal more accessible to residents.

Author details

Elsayed Elbeshbishy* and Frances Okoye
Civil Engineering Department, Ryerson University, Toronto, Canada

*Address all correspondence to: elsayed.elbeshbishy@ryerson.ca

References

[1] Inglezakis VJ, Moustakas K. Household hazardous waste management: A review. Journal of Environmental Management. 2015;**150**: 310-321

[2] Boadi KO, Kuitunen M. Environmental and health impacts of household solid waste handling and disposal practices in third world cites: The case of the Accra metropolitan area, Ghana. Journal of Environmental Health. 2005;**68**:32-37

[3] Wagner TP, Toews P, Bouvier R. Increasing diversion of household hazardous wastes and materials through mandatory retail take-back. Journal of Environmental Management. 2013;**123**: 88-97

[4] Slack RJ, Panagoula Z, Gronow JR, Voulvoulis N. Assessing quantities and disposal routes for household hazardous products in the United Kingdom. Environmental Science and Technology. 2005;**39**:1912-1919

[5] Statistics Canada. Households and the Environment. 2009. Available from: http://www.statcan.gc.ca/pub/11-526-x/11-526-x2011001-eng.htm

[6] Slack RJ, Gronow JR, Voulvoulis N. Hazardous components of household waste. Critical Reviews in Environmental Science and Technology. 2004;**34**:419-445

[7] Gendebien A, Leavens A, Blackmore K, Godley A, Lewin K, et al. Study on hazardous household waste (HHW) with a main emphasis on hazardous household chemicals (HHC). In: Report. 2002 http://ec.europa.eu/environment/waste/studies/pdf/household_report.pdf

[8] Rani B, Singh U, Yadav RK, Maheshwari R. Electronic waste: Its health hazards and management for sustainable era. Journal of Current Research in Science. 2013;**1**:157-163

[9] Thomas RHW. Factors influencing household hazardous waste disposal [master thesis]. Ottawa: Carlton University; 1998

[10] Bowen CF. Household Hazardous Products and Hazardous Waste: A Summary for Consumers. Pennsylvania, USA: The Pennsylvania State University; 1998

[11] Nightingale D, Donnette R. Household hazardous waste. In: Tchobanoglous G, Kreith F, editors. Handbook of Solid Waste Management. New York, USA: McGraw-Hill Professional Publishing; 2002

[12] Liu H, Liang Y, Zhang D, Wang C, Liang H, Cai H. Impact of MSW landfill on the environmental contamination of phthalate esters. Waste Management. 2010;**30**:1569-1576

[13] Ansari AA, Gill SS, Eutrophication KFA. Threat to aquatic ecosystems. In: Ansari AA, Gill SS, Lanza GR, Rast W, editors. Eutrophication: Causes, Consequences and Control. Netherlands: Springer; 2011. pp. 143-170. DOI: 10.1007/978-90-481-9625-8_7

[14] Xing GH, Chan JKY, Leung AOW, Wu SC, Wong MH. Environmental impact and human exposure to PCBs in Guiyu, an electronic waste recycling site in China. Environment International. 2009;**35**:76-82

[15] Leung A, Cai ZW, Wong MH. Environmental contamination from electronic waste recycling at Guiyu, Southeast China. Journal of Material Cycles and Waste Management. 2006;**8**: 21-33

[16] European Commission. Waste and Repealing Certain Directives. 2008. Available from: http://eur-lex.europa.eu/legal-content/EN/TXT/?uri=CELEX:32008L0098

[17] European Commission. Machinery and Amending Directive. 2006. Available from: http://eur-lex.europa.eu/legal-content/EN/TXT/?uri=celex%3A32006L0042

[18] Environmental Protection Agency. Title 40 Protection of Environment. 2016. Available from: http://www.ecfr.gov/cgi-bin/text-idx?tpl=/ecfrbrowse/Title40/40tab_02.tpl

[19] Gershman, Brickner & Bratton, Inc. Household hazardous, specialty, and electronic waste. In: Integrated Solid Waste Management Plan. City of Maui, Hawaii: County of Maui; 2009

[20] Environment and Climate Change Canada. Household Hazardous Waste Program. 2013. Available from: http://ec.gc.ca/gdd-mw/default.asp?lang=En&n=618F3E03-1